# PETROCHEMICAL PRODUCTION PROCESSES

# PETROCHEMICAL PRODUCTION PROCESSES

Fidel Russel

Oxford Book Company
Jaipur ▪ New Delhi

*Published By*
**Oxford** Book Company

**Regd. Off.** "Bony Residency", Gate No. 2
Opp. Tilak Public School, Vishweswariya Nagar,
Gopalpura Road, Jaipur - 302018, India
Ph: 0141 - 2761280, Telefax: 2761381
email - oxfordbookjpr@gmail.com
website - oxfordbookcompany.com

**Delhi Office:** 102, 1st Floor, Satyam House,
4327/3, Ansari Road, Darya Ganj,
New Delhi - 110002, India
Ph: 011-45652440

Distributed by
**KEYA INTERNATIONAL PRESS LLP**
E-mail: keyainternationalpressllp@gmail.com

First published in India, 2024



ISBN : 9789355240941



Typeset by Shivangi Computers, Jaipur

Printed and bound in India by Replika Press Pvt. Ltd.

# Contents

# Preface

The technique of hydraulic fracturing is used to increase or restore the rate at which fluids, such as oil or water, or natural gas can be produced from subterranean natural reservoirs, including unconventional reservoirs such as shale rock or coal beds. Hydraulic fracturing enables the production of natural gas and oil from rock formations deep below the earth's surface (generally 5,000-20,000 feet or 1,500-6,100 m). At such depth, there may not be sufficient porosity and permeability to allow natural gas and oil to flow from the rock into the wellbore at economic rates. Thus, creating conductive fractures in the rock is essential to extract gas from shale reservoirs because of the extremely low natural permeability of shale, which is measured in the microdarcy to nanodarcy range. Fractures provide a conductive path connecting a larger area of the reservoir to the well, thereby increasing the area from which natural gas and liquids can be recovered from the targeted formation.

A hydraulic fracture is formed by pumping the fracturing fluid into the wellbore at a rate sufficient to increase pressure downhole to exceed that of the fracture gradient of the rock. The rock cracks and the fracture fluid continues farther into the rock, extending the crack still farther, and so on. To keep this fracture open after the injection stops, a solid proppant, commonly a sieved round sand, is added to the fluid. The propped fracture is permeable enough to allow the flow of formation fluids to the well. Formation fluids include gas, oil, salt water, fresh water and fluids introduced to the formation during completion of the well during fracturing. The location of fracturing along the length of the borehole can be controlled by inserting composite plugs, also known as bridge plugs, above and below the region to be fractured. This allows a borehole to be progressively fractured along the length of the bore without fluid leaking through previously fractured regions. Fluid and proppant are introduced to the working region through piping in the upper plug. This method is commonly referred to as "plug and perf." Typically, hydraulic fracturing is performed in cased wellbores and the zones to be fractured are accessed by perforating the casing at those locations.

While hydraulic fracturing is many times performed in vertical wells, today it is more often performed in horizontal wells. Horizontal drilling involves wellbores where the terminal drillhole is completed as a 'lateral' that extends parallel with the rock layer containing the substance to be extracted. For example, laterals extend 1,500 to 5,000 feet in the Barnett Shale basin. In contrast, a vertical well only accesses the thickness of the rock layer, typically 50–300 feet. Horizontal drilling also reduces surface disruptions as fewer wells are required. Drilling a wellbore produces rock chips and fine rock particles that may enter cracks and pore space at the wellbore wall, reducing the permeability at and near the wellbore. This reduces flow into the borehole from the surrounding rock formation, and partially seals off the borehole from the surrounding rock. Hydraulic fracturing can be used to restore permeability. Hydraulic fracturing is commonly applied to wells drilled in low permeability reservoir rock. An estimated 90 percent of the natural gas wells in the United States use hydraulic fracturing to produce gas at economic rates.

This book contains advanced information about this subject. This book will be a boost for the learners and an essential subject manual. Designing of the book is such that the students will be benefited as far as the their knowledge and examination is concerned.

—***Fidel Russel***

# 1

# Acrylonitrile Butadiene Styrene

Acrylonitrile butadiene styrene (ABS) (chemical formula $(C_8H_8)_x \cdot (C_4H_6)_y \cdot (C_3H_3N)_z$) is a common thermoplastic. Its melting point is approximately 105 °C (221 °F).

It is a copolymer made by polymerizing styrene and acrylonitrile in the presence of polybutadiene. The proportions can vary from 15 to 35% acrylonitrile, 5 to 30% butadiene and 40 to 60% styrene. The result is a long chain of polybutadiene criss-crossed with shorter chains of poly(styrene-co-acrylonitrile). The nitrile groups from neighboring chains, being polar, attract each other and bind the chains together, making ABS stronger than pure polystyrene. The styrene gives the plastic a shiny, impervious surface. The butadiene, a rubbery substance, provides resilience even at low temperatures. For the majority of applications, ABS can be used between "25 and 60 °C (-13 and 140 °F) as its mechanical properties vary with temperature. The properties are created by rubber toughening, where fine particles of elastomer are distributed throughout the rigid matrix.

Production of 1 kg of ABS requires the equivalent of about 2 kg of petroleum for raw materials and energy. It can also be recycled.

### *Properties*

ABS is derived from acrylonitrile, butadiene, and styrene and carbon. Acrylonitrile is a synthetic monomer produced from propylene and ammonia; butadiene is a petroleum hydrocarbon obtained from the C4 fraction of steam cracking; styrene monomer is made by dehydrogenation of ethyl benzene — a hydrocarbon obtained in the reaction of ethylene and benzene.

The advantage of ABS is that this material combines the strength and rigidity of the acrylonitrile and styrene polymers with the toughness of the polybutadiene rubber. The most important mechanical properties of ABS are impact resistance and toughness. A variety of modifications can be made to improve impact resistance, toughness, and heat resistance. The impact resistance can be amplified by increasing the proportions of polybutadiene in relation to styrene and also acrylonitrile, although this causes changes in other properties. Impact resistance does not fall off rapidly at lower temperatures. Stability under load is excellent with limited loads. Thus, changing the proportions of its components ABS can be prepared in different grades. Two major categories could be ABS for extrusion and ABS for injection moulding, then high and medium impact resistance. Generally ABS would have useful characteristics within a temperature range from 10 to 80 °C (50 to 176 °F).

The final properties will be influenced to some extent by the conditions under which the material is processed to the final product. For example, molding at a high temperature improves the gloss and heat resistance of the product whereas the highest impact resistance and strength are obtained by molding at low temperature. Fibres (usually glass fibres) and additives can be mixed in the resin pellets to make the final product strong and raise the operating range to as high as 80 °C (176 °F). Pigments can also be added, as the raw material original colour is translucent ivory to white. The aging characteristics of the polymers are largely influenced by the polybutadiene content, and it is normal to include antioxidants in the composition. Other factors include exposure to ultraviolet radiation, for which additives are also available to protect against.

Even though ABS plastics are used largely for mechanical purposes, they also have electrical properties that are fairly constant over a wide range of frequencies. These properties are little affected by temperature and atmospheric humidity in the acceptable operating range of temperatures.

ABS polymers are resistant to aqueous acids, alkalis, concentrated hydrochloric and phosphoric acids, alcohols and animal, vegetable and mineral oils, but they are swollen by glacial acetic acid, carbon tetrachloride and aromatic hydrocarbons and are attacked by concentrated sulfuric and nitric acids. They are soluble in esters, ketones, ethylene dichloride and acetone.

While the cost of producing ABS is roughly twice the cost of producing polystyrene, it is considered superior for its hardness, gloss, toughness, and electrical insulation properties. ABS is flammable when it is exposed to high temperatures, such as a wood fire. It will melt then boil, at which point the vapours burst into intense, hot flames.

### *Applications*

ABS's light weight and ability to be injection molded and extruded make it useful in manufacturing products such as drain-waste-vent (DWV) pipe systems, musical instruments (recorders and plastic clarinets), golf club heads (due to its good shock absorbance), automotive trim components, automotive bumper bars, enclosures for electrical and electronic assemblies, protective headgear, whitewater canoes, buffer edging for furniture and joinery panels, luggage and protective carrying cases, small kitchen appliances, and toys, including Lego bricks.

ABS plastic ground down to an average diameter of less than 1 micrometer is used as the colorant in some tattoo inks. Tattoo inks that use ABS are extremely vivid. This vividness is the most obvious indicator that the ink contains ABS, as tattoo inks rarely list their ingredients.

## Synthetic Rubber

Synthetic rubber is any type of artificial elastomer, invariably a polymer. An elastomer is a material with the mechanical (or material) property that it can undergo much more elastic deformation under stress than most materials and still return to its previous size without permanent deformation. Synthetic rubber serves as a substitute for natural rubber in many cases, especially when improved material properties are required.

### *Comparison of Natural and Synthetic Rubber*

Natural rubber coming from latex is mostly polymerized isoprene with a small percentage of impurities in it. This limits the range of properties available to it. Also, there are limitations on the proportions of *cis* and *trans* double bonds resulting from methods of polymerizing natural latex. This also limits the range of properties available to natural rubber, although addition of sulfur and vulcanization are used to improve the properties.

Synthetic rubber can be made from the polymerization of a variety of monomers including isoprene (2-methyl-1,3-butadiene), 1,3-

butadiene, chloroprene (2-chloro-1,3-butadiene), and isobutylene (methylpropene) with a small percentage of isoprene for cross-linking. These and other monomers can be mixed in various desirable proportions to be copolymerized for a wide range of physical, mechanical, and chemical properties. The monomers can be produced pure and the addition of impurities or additives can be controlled by design to give optimal properties. Polymerization of pure monomers can be better controlled to give a desired proportion of *cis* and *trans* double bonds.

### History

In 1879, Bouchardat created one form of synthetic rubber, producing a polymer of isoprene in a laboratory.

The expanded use of motor vehicles, and particularly motor vehicle tires, starting in the 1890s, created increased demand for rubber.

In 1909, a team headed by Fritz Hofmann, working at the Bayer laboratory in Elberfeld, Germany, also succeeded in polymerizing methyl isoprene, the first synthetic rubber.

Scientists in England and Germany developed alternative methods for creating isoprene polymers from 1910–1912.

The Russian scientist Sergei Vasiljevich Lebedev created the first rubber polymer synthesized from butadiene in 1910. This form of synthetic rubber provided the basis for the first large-scale commercial production, which occurred during World War I as a result of shortages of natural rubber. This early form of synthetic rubber was again replaced with natural rubber after the war ended, but investigations of synthetic rubber continued. Russian American Ivan Ostromislensky did significant early research on synthetic rubber and a couple of monomers in the earlier 1900s.

Political problems that resulted from great fluctuations in the cost of natural rubber led to the enactment of the Stevenson Act in 1921. This act essentially created a cartel which supported rubber prices by regulating production, but insufficient supply, especially due to wartime shortages, also led to a search for alternative forms of synthetic rubber.

By 1925 the price of natural rubber had increased to the point that many companies were exploring methods of producing synthetic rubber to compete with natural rubber. In the United States, the investigation focused on different materials than in Europe, building on the early laboratory work of Nieuwland. Studies published in 1930 written independently by Lebedev, the American Wallace Carothers and the German scientist Hermann Staudinger led in 1931 to one of the first successful synthetic rubbers, known as neoprene, which was developed at DuPont under the direction of E.K. Bolton. Neoprene is highly resistant to heat and chemicals such as oil and gasoline, and is used in fuel hoses and as an insulating material in machinery.

The company Thiokol applied their name to a competing type of rubber based on ethylene dichloride which was commercially available in 1930.

The first rubber plant in Europe SK-1 (from Russian "Synthetic Kauchuk", Russian: ÑÊ-1) was established (Russia) by Sergei Lebedev in Yaroslavl under Stalin's First Five-Year Plan on July 7, 1932.

In 1935, German chemists synthesized the first of a series of synthetic rubbers known as Buna rubbers. These were copolymers, meaning the polymers were made up from two monomers in alternating sequence.

Other brands included *Koroseal*, which Waldo Semon developed in 1935, and *Sovprene*, which Russian researchers created in 1940.

### *World War II*

B.F. Goodrich Company scientist Waldo Semon developed a new and cheaper version of synthetic rubber known as Ameripol in 1940. Ameripol made synthetic rubber production much more cost effective, helping to meet the country's needs during World War II.

The production of synthetic rubber in the United States expanded greatly during World War II, since the Axis powers controlled nearly all the world's limited supplies of natural rubber by mid-1942 once Japan conquered Asia. Military trucks needed rubber for tires, and rubber was used in almost every other war machine. The U.S. government launched a major (and largely secret) effort to improve synthetic rubber production. A large team of chemists from many institutions were involved, including Calvin Souther Fuller of Bell Labs. The rubber designated GRS (Government Rubber Styrene), a copolymer of butadiene and styrene, was the basis for U.S. synthetic rubber production during World War II. By 1944, a total of 50 factories were manufacturing it, pouring out a volume of the material twice that of the world's natural rubber production before the beginning of the war. It still represents about half of total world production.

Operation Pointblank bombing targets of Nazi Germany included the Schkopau (50K tons/yr) plant and the Hüls synthetic rubber plant near Recklinghausen (30K, 17%), the Kölnische Gummifäden Fabrik tire and tube plant at Deutz on the east bank of the Rhine. The Ferrara, Italy, synthetic rubber factory (near a river bridge) was bombed August 23, 1944. Three other synthetic rubber facilities were at Ludwigshafen/Oppau (15K), Hanover/Limmer (reclamation, 20K), and Leverkusen (5K). A synthetic rubber plant at Oœwiêcim, Poland, was under construction on March 5, 1944.

### Post-war

Solid-fuel rockets during World War II used nitrocellulose explosives for propellants, but it was impractical and dangerous to make such rockets very large. During the war, California Institute of Technology (Caltech) researchers came up with a new solid fuel based on asphalt fuel mixed with an oxidizer (such as potassium or ammonium perchlorate), and aluminium powder. This new solid fuel burned more slowly and evenly than nitrocellulose explosives, and was much less dangerous to store and use, but it tended to flow slowly out of the rocket in storage and the rockets using it had to be stockpiled nose down.

After the war, Caltech researchers began to investigate the use of synthetic rubbers to replace asphalt in their solid fuel rocket motors. By the mid-1950s, large missiles were being built using solid fuels based on synthetic rubber, mixed with ammonium perchlorate and high proportions of aluminium powder. Such solid fuels could be cast into large, uniform blocks that had no cracks or other defects that would cause non-uniform burning. Ultimately, all large solid-fuel military rockets and missiles would use synthetic-rubber-based solid fuels, and they would also play a significant part in the civilian space effort.

Additional refinements to the process of creating synthetic rubber continued after the war. The chemical synthesis of isoprene accelerated the reduced need for natural rubber, and the peacetime quantity of synthetic rubber exceeded the production of natural rubber by the early 1960s.

Nowadays synthetic rubber is used a great deal in printing textile. In this case it is called rubber paste. In most cases titanium dioxide is used with copolymerization and volatile matter in producing such synthetic rubber for textile use. Moreover this kind of preparation can be considered to be the pigment preparation based on titanium dioxide. In 2005, close to 21 million tons of rubber were produced of which around 58% was synthetic.

*Table 1 : Table of common synthetic rubbers*

| ***ISO 1629 Code*** | ***Technical Name*** | ***Common Names*** |
|---|---|---|
| ACM | Polyacrylate Rubber | |
| AEM | Ethylene-acrylate Rubber | |
| AU | Polyester Urethane | |
| BIIR | Bromo Isobutylene Isoprene | Bromobutyl |
| BR | Polybutadiene | Buna CB |
| CIIR | Chloro Isobutylene Isoprene | Chlorobutyl, Butyl |
| CR | Polychloroprene | Chloroprene, Neoprene |
| CSM | Chlorosulphonated Polyethylene | Hypalon |
| ECO | Epichlorohydrin | ECO, Epichlorohydrin, Epichlore, Epichloridrine, Herclor, Hydrin |
| EP | Ethylene Propylene | |
| EPDM | Ethylene Propylene Diene Monomer | EPDM, Nordel |
| EU | Polyether Urethane | |
| FFKM | Perfluorocarbon Rubber | |
| FKM | Fluoronated Hydrocarbon | Viton, Kalrez, Fluorel, Chemraz |
| FMQ | Fluoro Silicone | FMQ, Silicone Rubber |
| FPM | Fluorocarbon Rubber | |
| HNBR | Hydrogenated Nitrile Butadiene | HNBR |
| IR | Polyisoprene | (Synthetic) Natural Rubber |
| IIR | Isobutylene Isoprene Butyl | Butyl |
| NBR | Acrylonitrile Butadiene | NBR, Nitrile rubber, Perbunan, Buna-N |
| PU | Polyurethane | PU, Polyurethane |
| SBR | Styrene Butadiene | SBR, Buna-S, GRS, Buna VSL, Buna SE |
| SEBS | Styrene Ethylene Butylene Styrene Copolymer | SEBS Rubber |
| SI | Polysiloxane | Silicone Rubber |

| | | |
|---|---|---|
| VMQ | Vinyl Methyl Silicone | Silicone Rubber |
| XNBR | Acrylonitrile Butadiene Carboxy Monomer | XNBR, Carboxylated Nitrile |
| XSBR | Styrene Butadiene Carboxy Monomer | |
| YBPO | Thermoplastic Polyether-ester | |
| YSBR | Styrene Butadiene Block Copolymer | |
| YXSBR | Styrene Butadiene Carboxy Block Copolymer | |

In addition the term *gum rubber* is sometimes used to describe the tree-derived *natural rubber* (code NR), and to distinguish it from synthetic *natural rubber* (code IR).

## Butyl Rubber

Butyl rubber is a synthetic rubber, a copolymer of isobutylene with isoprene. The abbreviation IIR stands for Isobutylene Isoprene Rubber. Polyisobutylene, also known as "PIB" or polyisobutene, $(C_4H_8)_n$, is the homopolymer of isobutylene, or 2-methyl-1-propene, on which butyl rubber is based. Butyl rubber is produced by polymerization of about 98% of isobutylene with about 2% of isoprene. Structurally, polyisobutylene resembles polypropylene, having two methyl groups substituted on every other carbon atom. Polyisobutylene is a colorless to light yellow viscoelastic material. It is generally odorless and tasteless, though it may exhibit a slight characteristic odor.

Butyl rubber has excellent impermeability, and the long polyisobutylene segments of its polymer chains give it good flex properties.

The formula for PIB is: $\text{-(-CH}_2\text{-C(CH}_3)_2\text{-)}_n\text{-}$

The formula for IIR is:

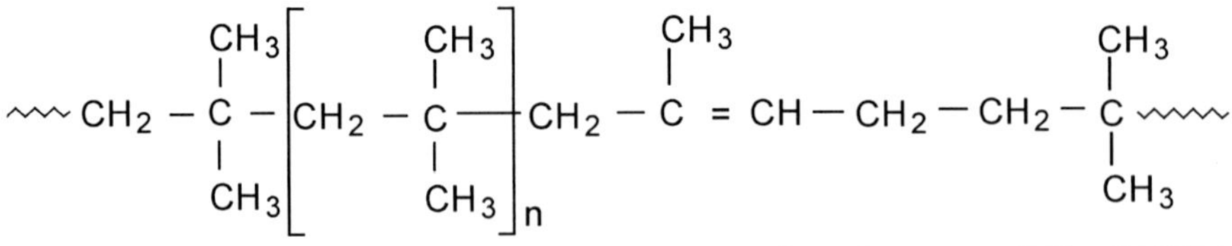

It can be made from the monomer isobutylene or $CH_2{=}C(CH_3)_2$ only via cationic addition polymerization.

A synthetic rubber, or elastomer, butyl rubber is impermeable to air and used in many applications requiring an airtight rubber.

Polyisobutylene and butyl rubber are used in the manufacture of adhesives, agricultural chemicals, fibre optic compounds, ball bladders, caulks and sealants, cling film, electrical fluids, lubricants (2 cycle engine oil), paper and pulp, personal care products, pigment concentrates, for rubber and polymer modification, for protecting and sealing certain equipment for use in areas where chemical weapons are present, as a gasoline/diesel fuel additive, and even in chewing gum. The first major application of butyl rubber was tire inner tubes. This remains an important segment of its market even today.

### History

Polyisobutylene was first developed by the BASF unit of IG Farben in 1931 and sold under the trade name Oppanol B. It was later developed into butyl rubber in 1937, by researchers William J. Sparks and Robert M. Thomas, at Standard Oil of New Jersey's Linden, N.J., laboratory. Today, the majority of the global supply of butyl rubber is produced by just two companies, ExxonMobil, the successor to Standard Oil, and LANXESS, a spinoff from Bayer. Texas Petrochemicals (TPC Group) is the largest manufacturer of PIB in North America.

In 1950s and 1960s, halogenated butyl rubber (halobutyl) was developed, in its chlorinated (chlorobutyl) and brominated (bromobutyl) variants, providing significantly higher curing rates and allowing covulcanization with such other rubbers as natural rubber and styrene-butadiene rubber. Halobutyl is today the most important material for the inner linings of tubeless tires.

### Uses

### Fuel and Lubricant Additive

Polyisobutylene (in the form of polyisobutylene succinimide, PIBSI) has interesting properties when used as an additive in lubricating oils and motor fuels. Polyisobutylene added in small amounts to the lubricating oils used in machining results in a significant reduction in the generation of oil mist and thus reduces the operator's inhalation of oil mist. It is also used to clean up waterborne oil spills as part of the commercial product Elastol. When added to crude oil it increases the oil's viscoelasticity when pulled, causing the oil to resist breakup when it is vacuumed from the surface of the water.

As a fuel additive, polyisobutylene has detergent properties. When added to diesel fuel, it resists fouling of fuel injectors, leading to reduced hydrocarbon and particulate emissions. It is blended with other detergents and additives to make a "detergent package" that is added to gasoline and diesel fuel to resist buildup of deposits and engine knock.

Polyisobutylene is used in some formulations as a thickening agent.

### *Sporting Equipment*

Butyl rubber is used for the bladders in basketballs, footballs, soccer balls and other inflatable balls to provide a tough, airtight inner compartment.

### *Roof Repair*

Butyl rubber sealant is used for rubber roof repair and for maintenance of roof membranes (especially around the edges). It is important to have the roof membrane fixed, as a lot of fixtures (i.e., air conditioner vents, plumbing and other pipes, etc.) can considerably loosen it.

Rubber roofing typically refers to a specific type of roofing materials that are made of ethylene propylene diene monomers (EPDM). It is crucial to the integrity of such roofs to avoid using harsh abrasive materials and petroleum-based solvents for their maintenance.

Polyester fabric laminated to butyl rubber binder provides a single-sided waterproof tape that can be used on metal, PVC, and cement joints. It is ideal for repairing and waterproofing metal roofs.

### *Gas Masks and Chemical Agent Protection*

Butyl rubber is one of the most robust elastomers when subjected to chemical warfare agents and decontamination materials. It is a harder and less porous material than other elastomers, such as natural rubber or silicone, but still has enough elasticity to form an airtight seal.

While butyl rubber will break down when exposed to agents such as $NH_3$ (ammonia) or certain solvents, it breaks down more slowly than comparable elastomers. It is therefore used to create seals in gas masks and other protective clothing.

### *Chewing Gum*

Many varieties of chewing gum utilize food-grade butyl rubber as the central gum base.

## 1,3-Butadiene

1,3-Butadiene is a simple conjugated diene with the formula $C_4H_6$. It is an important industrial chemical used as a monomer in the production of synthetic rubber. When the word butadiene is used, most of the time it refers to 1,3-butadiene.

The name butadiene can also refer to the isomer, 1,2-butadiene, which is a cumulated diene. However, this allene is difficult to prepare and has no industrial significance.

### History

In 1863, a French chemist isolated a previously unknown hydrocarbon from the pyrolysis of amyl alcohol. This hydrocarbon was identified as butadiene in 1886, after Henry Edward Armstrong isolated it from among the pyrolysis products of petroleum. In 1910, the Russian chemist Sergei Lebedev polymerized butadiene, and obtained a material with rubber-like properties. This polymer was, however, too soft to replace natural rubber in many roles, especially automobile tires.

The butadiene industry originated in the years leading up to World War II. Many of the belligerent nations realized that in the event of war, they could be cut off from rubber plantations controlled by the British Empire, and sought to remove their dependence on natural rubber. In 1929, Eduard Tschunker and Walter Bock, working for IG Farben in Germany, made a copolymer of styrene and butadiene that could be used in automobile tires. Worldwide production quickly ensued, with butadiene being produced from grain alcohol in the Soviet Union and the United States and from coal-derived acetylene in Germany.

### Production

### Extraction from C4 Hydrocarbons

In the United States, western Europe, and Japan, butadiene is produced as a byproduct of the steam cracking process used to produce ethylene and other olefins. When mixed with steam and briefly heated to very high temperatures (often over 900 °C), aliphatic hydrocarbons give up hydrogen to produce a complex mixture of unsaturated hydrocarbons, including butadiene. The quantity of butadiene produced depends on the hydrocarbons used as feed. Light feeds, such as ethane, give primarily ethylene when cracked, but heavier feeds favour the formation of heavier olefins, butadiene, and aromatic hydrocarbons.

Butadiene is typically isolated from the other four-carbon hydrocarbons produced in steam cracking by extraction into a polar aprotic solvent such as acetonitrile, N-methylpyrrolidone, furfural, or dimethylformamide, from which it is then stripped by distillation.

### *From Dehydrogenation of n-Butane*

Butadiene can also be produced by the catalytic dehydrogenation of normal butane. The first such post-war commercial plant, producing 65,000 tons per year of butadiene, began operations in 1957 in Houston, Texas. Prior to that, in the 1940s the U. S. War Department constructed several much larger plants in Borger, TX, Toledo, OH, and El Segundo, CA to produce synthetic rubber for the war effort as part of the United States Synthetic Rubber Program.

### *From Ethanol*

. In other parts of the world, including eastern Europe, China, and India, butadiene is also produced from ethanol. While not competitive with steam cracking for producing large volumes of butadiene, lower capital costs make production from ethanol a viable option for smaller-capacity plants. Two processes are in use.

In the single-step process developed by Sergei Lebedev, ethanol is converted to butadiene, hydrogen, and water at 400–450 °C over any of a variety of metal oxide catalysts:

$$2\ CH_3CH_2OH \rightarrow CH_2{=}CH\text{-}CH{=}CH_2 + 2\ H_2O + H_2$$

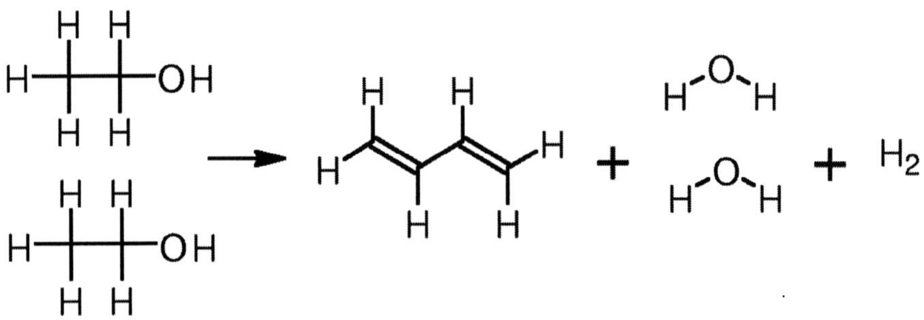

This process was the basis for the Soviet Union's synthetic rubber industry during and after World War II, and it remains in limited use in Russia and other parts of eastern Europe.

In the other, two-step process, developed by the Russian chemist Ivan Ostromislensky, ethanol is oxidized to acetaldehyde, which reacts with additional ethanol over a tantalum-promoted porous silica catalyst at 325–350 °C to yield butadiene:

$$CH_3CH_2OH + CH_3CHO \rightarrow CH_2{=}CH\text{-}CH{=}CH_2 + 2\ H_2O$$

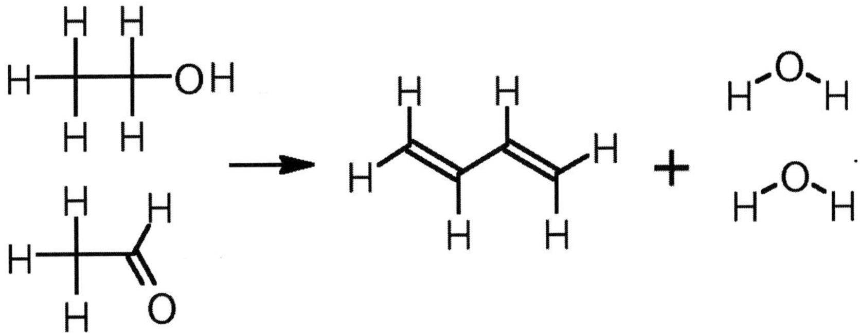

This process was also used in the United States to produce government rubber during World War II, though it was not preferred because it is less economical than the butane or butene routes for the large volumes needed. It remains in use today in China and India.

### From Butenes

1,3-Butadiene can also be produced by catalytic dehydrogenation of normal butenes. This method was also used by the United States Synthetic Rubber Program (USSRP) during World War II. The process was much more economical than the alcohol route but competed with aviation gasoline for available butene molecules. The USSRP constructed several plants in Baton Rouge and Lake Charles, LA; Houston, Baytown, and Port Neches, TX; and Torrance, CA. In the 1960s, a Houston company known as "PetroTex" patented a process to produce butadiene from normal butenes by oxidative dehydrogenation using a proprietary catalyst. It is thought to be no longer practiced commercially.

### For Laboratory Use

1,3-Butadiene is inconvenient for laboratory use because it is a flammable gas subject to polymerization on storage. 3-Butadiene cyclic sulfone (sulfolene) is a convenient solid storable source for 1,3-butadiene for many laboratory purposes when the generation of sulfur dioxide byproduct in the reaction mixture is not objectionable.

### Uses

Most butadiene is polymerized to produce synthetic rubber. While polybutadiene itself is a very soft, almost liquid material, copolymers prepared from mixtures of butadiene with styrene and/or acrylonitrile, such as acrylonitrile butadiene styrene (ABS), acrylonitrile butadiene (NBR) and styrene-butadiene (SBR) are tough and elastic. SBR is the material most commonly used for the production of automobile tires.

Smaller amounts of butadiene are used to make the nylon intermediate, adiponitrile, by the addition of a molecule of hydrogen cyanide to each of the double bonds in a process called hydrocyanation developed by DuPont. Other synthetic rubber materials such as chloroprene, and the solvent sulfolane are also manufactured from butadiene. Butadiene is used in the industrial production of 4-vinylcyclohexene via a Diels Alder dimerization reaction and the vinylcyclohexene is a common impurity found in butadiene upon storage. Cyclooctadiene and cyclododecatriene are produced via nickel- or titanium-catalysed dimerization and trimerization reactions, respectively. Butadiene is also useful in the synthesis of cycloalkanes and cycloalkenes, as it reacts with double and triple carbon-carbon bonds through the Diels-Alder reaction.

### *Environmental Health and Safety*

Acute exposure results in irritation of the mucous membranes, Higher levels can result in neurological effects such as blurred vision, fatigue, headache and vertigo. Exposure to the skin can lead to frostbite.

Long-term exposure has been associated with cardiovascular disease, There is a consistent association with leukemia, and weaker association with other cancers.

1,3 Butadiene is listed as a known carcinogen by the Agency for Toxic Substances Disease Registry and the US EPA. The American Council for Governmental Industrial Hygienists (ACGIH) lists the chemical as a suspected carcinogen. The Natural Resource Defence Council (NRDC) lists some disease clusters that are suspected to be associated with this chemical.

1,3-Butadiene is also a suspected human teratogen. Prolonged and excessive exposure can affect many areas in the human body; blood, brain, eye, heart, kidney, lung, nose and throat have all been shown to react to the presence of excessive 1,3-Butadiene. Animal data suggest that women have a higher sensitivity to possible carcinogenic effects of butadiene over men when exposed to the chemical. This may be due to estrogen receptor impacts. While these data reveal important implications to the risks of human exposure to butadiene, more data are necessary to draw conclusive risk assessments. There is also a lack of human data for the effects of butadiene on reproductive and development shown to occur in mice, but animal studies have shown breathing butadiene during pregnancy can increase the number of birth defects, and humans have the same hormone systems as animals.

Storage of butadiene as a compressed, liquified gas carries a specific and unusual hazard. Over time, polymerization can begin, creating a crust of solidified material (popcorn polymer, named for its appearance) inside the cylinder. If the cylinder is then disturbed, the crust can contact the liquid and initiate an auto-catalytic polymerization. The heat released accelerates the reaction, possibly leading to cylinder rupture. Inhibitors are typically added to reduce this hazard, but butadiene cylinders should still be considered short-shelf life items.

## Diene

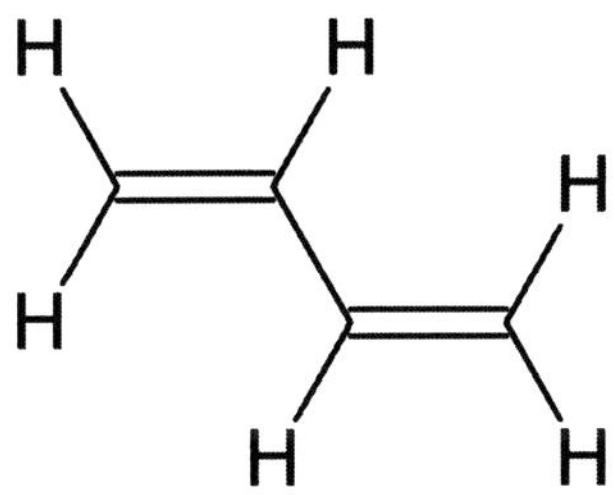

*1,3-butadiene*

In organic chemistry a diene is a hydrocarbon that contains two carbon double bonds. Conjugated dienes are functional groups, with a general formula of $C_nH_{2n-2}$. Dienes and alkynes are functional isomers. Dienes occur occasionally in nature but are widely used in the polymer industry.

### *Classes*

Dienes can be divided into three classes, depending on the relative location of the double bonds:

1. Cumulated dienes have the double bonds sharing a common atom as in a group of compounds called allenes.
2. Conjugated dienes have conjugated double bonds separated by one single bond.

3. Unconjugated dienes have the double bonds separated by two or more single bonds. They are usually less stable than isomeric conjugated dienes. This can also be known as an isolated diene.

Compounds that contain more than two double bonds are called polyenes. Polyenes and dienes, share many of their properties.

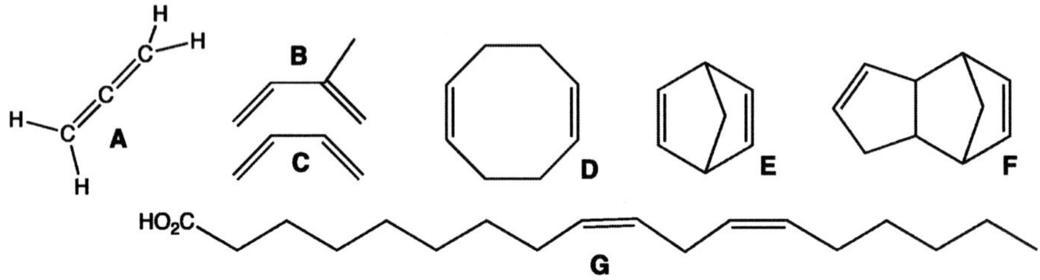


Some dienes: A: 1,2-Propadiene, also known as allene, is the simplest cumulated diene. B: Isoprene, also known as 2-methyl-1,3-butadiene, the precursor to natural rubber. C: 1,3-Butadiene, a precursor to synthetic polymers.

D: 1,5-Cyclooctadiene, an unconjugated diene (notice that each double bond is two carbons away from the other). E: Norbornadiene, a strained bicyclic and unconjugated diene. F: Dicyclopentadiene. G: Linoleic acid, a fatty acid that is required in the human diet.

### *Synthesis of Dienes*

On an industrial scale, butadiene is prepared by thermal cracking of butanes. In a similarly non-selective process, dicyclopentadiene is obtained from coal tars.

In the laboratory, more directed and more delicate processes are employed such as dehydrohalogenations and condensations. Myriad methods have been developed, such as the Whiting reaction. Families of nonconjugated dienes are derived from the oligomerization and dimerization of conjugated dienes. For example, 1,5-cyclooctadiene and vinylcyclohexene are produced by dimerization of 1,3-butadiene.

## Reactivity and Uses

### *Polymerization*

The most heavily practiced reaction of alkenes, dienes included, is polymerization. Butadiene is a precursor to rubber used in tires, and isoprene is the precursor to natural rubber. Chloroprene is a related but synthetic monomer.

### Cycloadditions

An important reaction for conjugated dienes is the Diels-Alder reaction. Many specialized dienes have been developed to exploit this reactivity for the synthesis of natural products, e.g. Danishefsky's diene.

### Other Addition Reactions

Conjugated dienes add reagents such as bromine and hydrogen by both 1,2-addition and 1,4-addition pathways. Addition of polar reagents can generate complex architectures:

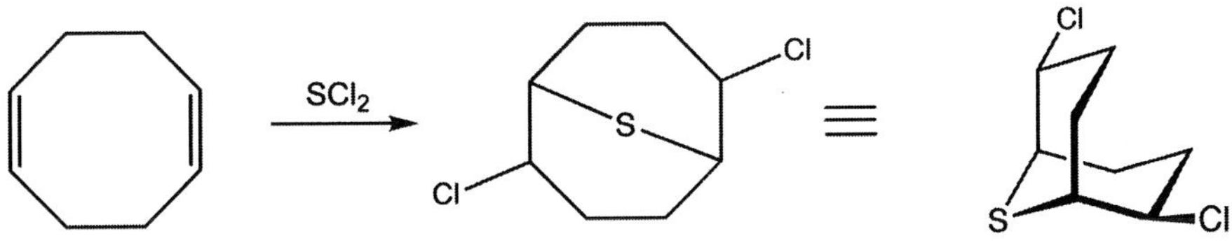


### Acidity

Conjugated and 1,4-dienes generally are somewhat acidic since deprotonation of both classes gives pentadienyl anions. The acidifying effect of the diene is very pronounced in cyclopentadiene.

### As Ligands

Dienes are widely used chelating ligands in organometallic chemistry. In some cases they serve as placeholder ligands, being removed during a catalytic cycle. For example, the cyclooctadiene ("cod") ligands in bis(cyclooctadiene)nickel(0) are labile. In some cases, dienes are spectator ligands, remaining coordinated throughout a catalytic cycle and influencing the product distributions. Chiral dienes have also been described.

## Elastomer

An elastomer is a polymer with the property of viscoelasticity (colloquially "elasticity"), generally having notably low Young's modulus and high yield strain compared with other materials. The term, which is derived from *elastic polymer*, is often used interchangeably with the term rubber, although the latter is preferred when referring to vulcanisates. Each of the monomers which link to form the polymer is usually made of carbon, hydrogen, oxygen and/or silicon. Elastomers are amorphous polymers existing above their glass transition temperature, so that considerable segmental motion is possible. At ambient temperatures, rubbers are thus relatively soft (E~3MPa) and deformable. Their primary uses are for seals, adhesives and molded flexible parts.

### Background

Elastomers are usually thermosets (requiring vulcanization) but may also be thermoplastic. The long polymer chains cross-link during curing, i.e., vulcanizing. The molecular structure of elastomers can be imagined as a 'spaghetti and meatball' structure, with the meatballs signifying cross-links. The elasticity is derived from the ability of the long chains to reconfigure themselves to distribute an applied stress. The covalent cross-linkages ensure that the elastomer will return to its original configuration when the stress is removed. As a result of this extreme flexibility, elastomers can reversibly extend from 5-700%, depending on the specific material. Without the cross-linkages or with short, uneasily reconfigured chains, the applied stress would result in a permanent deformation.

Temperature effects are also present in the demonstrated elasticity of a polymer. Elastomers that have cooled to a glassy or crystalline phase will have less mobile chains, and consequentially less elasticity, than those manipulated at temperatures higher than the glass transition temperature of the polymer.

It is also possible for a polymer to exhibit elasticity that is not due to covalent cross-links, but instead for thermodynamic reasons.

### Examples of Elastomers

Unsaturated rubbers that can be cured by sulfur vulcanization:

- Natural polyisoprene: cis-1,4-polyisoprene natural rubber (NR) and trans-1,4-polyisoprene gutta-percha
- Synthetic polyisoprene (IR for Isoprene Rubber)
- Polybutadiene (BR for Butadiene Rubber)
- Chloroprene rubber (CR), polychloroprene, Neoprene, Baypren etc.
- Butyl rubber (copolymer of isobutylene and isoprene, IIR)
    - Halogenated butyl rubbers (chloro butyl rubber: CIIR; bromo butyl rubber: BIIR)
- Styrene-butadiene Rubber (copolymer of styrene and butadiene, SBR)

- Nitrile rubber (copolymer of butadiene and acrylonitrile, NBR), also called Buna N rubbers
    - o Hydrogenated Nitrile Rubbers (HNBR) Therban and Zetpol.

      (Unsaturated rubbers can also be cured by non-sulfur vulcanization if desired).

Saturated rubbers that cannot be cured by sulfur vulcanization:

- EPM (ethylene propylene rubber, a copolymer of ethylene and propylene) and EPDM rubber (ethylene propylene diene rubber, a terpolymer of ethylene, propylene and a diene-component)
- Epichlorohydrin rubber (ECO)
- Polyacrylic rubber (ACM, ABR)
- Silicone rubber (SI, Q, VMQ)
- Fluorosilicone Rubber (FVMQ)
- Fluoroelastomers (FKM, and FEPM) Viton, Tecnoflon, Fluorel, Aflas and Dai-El
- Perfluoroelastomers (FFKM) Tecnoflon PFR, Kalrez, Chemraz, Perlast
- Polyether block amides (PEBA)
- Chlorosulfonated polyethylene (CSM), (Hypalon)
- Ethylene-vinyl acetate (EVA).

One should go through all differentiation while editing between Plastics and articles thereof and Rubber and articles thereof."

Various other types of elastomers:

- Thermoplastic elastomers (TPE)
- The proteins resilin and elastin
- Polysulfide rubber.

## Polybutadiene

Polybutadiene is a synthetic rubber that is a polymer formed from the polymerization process of the monomer 1,3-butadiene.

It has a high resistance to wear and is used especially in the manufacture of tires, which consumes about 70% of the production. Another 25% is used as an additive to improve the mechanical strength of plastics such as polystyrene and acrylonitrile butadiene styrene (ABS). It is also used to manufacture golf balls, various elastic objects and to coat or encapsulate electronic assemblies, offering extremely high electrical resistivity. It exhibits a recovery of 80% after stress is applied, a value only exceeded by elastin and resilin.

The butadiene molecule can be polymerized in three different ways, giving rise to three isomers called *cis*, *trans* and *vinyl*. The properties of polybutadiene are different depending on the proportion of these isomers. For example, polybutadiene called "high *cis*" has a high elasticity and is very popular while the so-called "high *trans*" is a plastic crystal without any useful application.

The annual production of polybutadiene is 2.1 million tons (2000). This makes it the second most produced synthetic rubber by volume, behind styrene-butadiene rubber (SBR).

Butadiene was first polymerized in 1910 by the Russian chemist Sergei Vasilyevich Lebedev. Due to his discoveries, the Soviet Union became the first country to achieve a substantial industrial production of the material in the late 1930's. In the same time, other powerful countries such as Germany and the United States decided to develop SBR as an alternative to natural rubber.

In the mid 1950's there were major advances in the field of catalysts that led to the development of an improved versions of polybutadiene. The leading manufacturers of tires and some petrochemical companies began to build polybutadiene plants on all continents; the boom lasted until the 1973 oil crisis. Since then, the growth rate of the production has been more modest, focused mainly to the Far East.

Other names recommended by IUPAC are: poly (buta-1,3-diene) and poly (but-1-ene-1,4-diyl).

### History

The Russian chemist Sergei Vasilyevich Lebedev was the first to polymerize butadiene in 1910. In 1926 he invented a process for manufacturing butadiene from ethanol, and in 1928, developed a method for producing polybutadiene using sodium as a catalyst.

The government of the Soviet Union strived to use polybutadiene as an alternative to natural rubber and built the first pilot plant in 1930, using ethanol produced from potatoes. The experiment was a success and in 1936 the Soviet Union built the world's first polybutadiene plant in which the butadiene was obtained from petroleum. By 1940, the Soviet Union was by far the largest producer of polybutadiene with 50,000 tons per year.

In Germany, scientists from Bayer (at the time a part of the conglomerate IG Farben) reproduced Lebedev's processes of producing polybutadiene by using sodium as a catalyst. For this, they used the trade name Buna, derived from Bu for butadiene, Na for sodium (natrium in Latin, Natrium in German). They discovered that the addition of styrene to the process resulted in better properties, and thus opted for this route. They had invented styrene-butadiene, which was named Buna-S (S for styrene).

Although the Goodrich Corporation had successfully developed a process for producing polybutadiene in 1939, the government of the United States opted for the use of Buna-S to develop its synthetic rubber industry after its entry into the World War II, using patents of IG Farben obtained via Standard Oil. Because of this, there was little industrial production of polybutadiene in America during this time.

After the war, the production of synthetic rubber was in decline due to the decrease in demand when natural rubber re-emerged in the market. However, interest was renewed in the mid 1950's after the discovery of the Ziegler-Natta catalyst. This method proved to be much better for tire manufacturing than the old sodium polybutadiene. The following year, Firestone Tire and Rubber Company was first to produce low *cis* polybutadiene using butyllithium as a catalyst.

The relatively high production costs were a hinderance to commercial development until 1960 when production on a commercial scale emerged after sufficient technological advances. Tire manufacturers like Goodyear Tire and Rubber Company and Goodrich were the first to produce plants for high *cis* polybutadiene, this was followed by oil companies like Shell and chemical manufacturers such as Bayer.

Initially, with plants built in the United States and France, Firestone had monopoly on low *cis* polybutadiene, licensing it to plants in Japan and the United Kingdom. In 1965, the Japanese JSR Corporation developed its own low *cis* process and began licensing it during the next decade.

The 1973 oil crisis marked a halt to the growth of synthetic rubber production; the expansion of existing plants almost ceased for a few years. Since then, the construction of new plants have been mainly focused to industrializing countries in the Far East (such as South Korea, Taiwan, Thailand and China), while Western countries have chosen to increase the capacity of existing plants. In 1987, Bayer started to use neodymium to catalyze polybutadiene; this move was soon followed by other manufacturers such as EniChem (1993) and Petroflex (2002).

In the early 2000s, the synthetic rubber industry was once again hit by one its periodic crises. The world's largest producer of polybutadiene, Bayer, went through major restructurings as they were troubled by financial losses; between 2002 and 2005 they closed its cobalt-polybutadiene plants in Sarnia (Canada) and Marl (Germany), trasferring their production to neodymium plants in Port Jérôme (France) and Orange (USA). During the same time, the synthetic rubber business was transferred from Bayer to Lanxess, a company founded in 2004 when Bayer spun off its chemicals operations and parts of its polymer activities.

### Polymerization of Butadiene

1,3-Butadiene is an organic compound that is a rather simple conjugated diene hydrocarbon; the chemical structure is shown as a reactant in the diagram below. A hydrocarbon diene molecule has two carbon-carbon double bonds (i. e. between two sets of carbon atoms). Polybutadiene can be formed from many 1,3-butadiene monomers undergoing free radical polymerization to make a much longer polymer chain molecule.

A chain propagating step in this chemical reaction involves a free radical near the end of a growing polymer chain forming a covalent bond with the #1 carbon in a 1,3-butadiene monomer molecule being added, resulting in a polymer chain intermediate with a substituted allyl free radical at the end of the chain. This allyl free radical, formed from the butadiene just added, can further bond to another monomer molecule at either the #2 or #4 carbons of the previous butadiene monomer. Most of the time, the new monomer bonds to the #4 or terminal carbon of the previous butadiene, resulting in a 1,4-addition of the previous butadiene unit. In a 1,4-addition, the two double bonds of the previous butadiene unit are turned into single bonds and a new double bond is formed between the #2 and #3 carbons. This new double bond may have either a *cis* or a *trans* configuration. A smaller fraction of the time (perhaps 20%), the new monomer bonds to the #2 carbon of the previous butadiene, resulting in a 1,2-addition of the previous butadiene unit. The double bond between the #1 and #2 carbons turns into a single bond in the previous butadiene unit, and the double bond between the #3 and #4 carbons remains intact in a short vinyl side group available for branching or cross-linking. *Cis* or *trans* configurations are not applicable in 1,2-additions of butadiene.

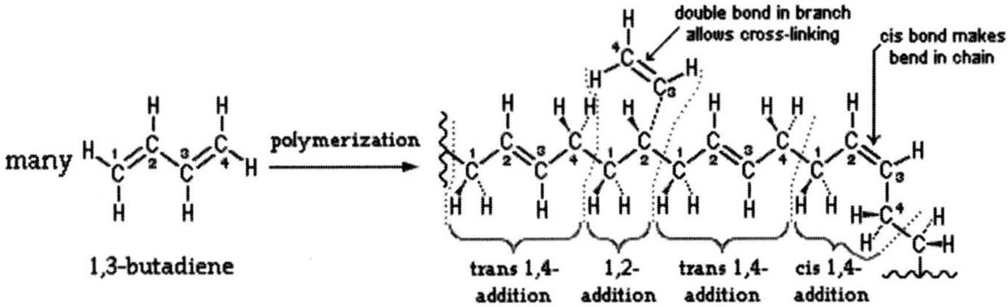


The *trans* double bonds formed during polymerization allow the polymer chain to stay rather straight, allowing sections of polymer chains to line up against each other and effectively form microcrystalline regions in the material. The *cis* double bonds cause a bend in the polymer chain, preventing polymer chains from lining up and forming crystalline regions and resulting in larger regions of amorphous polymer. It has been found that a substantial percentage of *cis* double bond configurations in the polymer will result in a material with flexible elastomer (rubber-like) qualities. In free radical polymerization, both *cis* and *trans* double bonds will form in percentages that depend on temperature. There are different catalysts available that can result in polymerization either in the *cis* or the *trans* configurations.

## Types

Depending on the catalyst used in the production, the type of polybutadiene will be different, each with specific properties.

*Table 2 : Typical composition of polybutadiene based on the catalyst used*

| | ***cis* (%)** | ***trans* (%)** | ***vinyl* (%)** |
|---|---|---|---|
| Neodymium | 98 | 1 | 1 |
| Cobalt | 96 | 2 | 2 |
| Nickel | 96 | 3 | 1 |
| Titanium | 93 | 3 | 4 |
| Lithium | 10 to 30 | 20 to 60 | 10 to 70 |

## High cis Polybutadiene

This type is characterized by a high proportion of *cis* (typically over 92%) and a small proportion of vinyl (less than 4%). It is manufactured using Ziegler-Natta catalysts based on transition metals. Depending on the metal used, the properties vary slightly.

Using cobalt gives branched molecules, resulting in a low viscosity material that is ease of use, but its mechanical strength is relatively low. Nodymium gives the most linear structure (and therefore higher mechanical strength) and a higher percentage of 98% *cis*. Other less used catalysts include nickel and titanium.

### Low cis Polybutadiene

Using an alkyllithium (e.g. butyllithium) as the catalyst produces a polybutadiene called "low *cis*" which typically contains 36% *cis*, 54% *trans* and 10% vinyl.

Because of its high liquid-glass transition, low *cis* polybutadiene is not used in tire manufacturing, but it can be advantagously used as an additive in plastics due to its low contents of gels.

### High Vinyl Polybutadiene

In 1980, researchers from Zeon discovered that high-vinyl polybutadiene (over 70%), despite having a high liquid-glass transition, could be advantageously used in combination with high *cis* in tires. This material is produced with an alkyllithium catalyst. In addition to the Japanese company Zeon, the American company Firestone produces high-vinyl polybutadiene as well.

JSR Corporation markets a type of polybutadiene with 90% vinyl, giving it the properties of an elastomeric thermoplastic: elastic at room temperature but a fluid at high temperatures, which makes it possible to process it using injection molding.

### High Trans Polybutadiene

Polybutadiene can be produced with more than 90% *trans* using catalysts similar to those of high *cis*: neodymium, lanthanum, nickel. This material is a plastic crystal (i.e. not an elastomer) which melts at about 80 °C. It was formerly used for the outer layer of golf balls. Today it is only used industrially, but companies like Ube are investigating other possible applications.

### Metallocene Polybutadiene

The use of metallocene catalysts to polymerize butadiene is being explored by Japanese researchers. The benefits seem to be a higher degree of control both in the distribution of molecular mass and the proportion of cis/trans/vinyl. As of 2006, no manufacturer produces "metallocene polybutadiene" on a commercial basis.

### Properties

Polybutadiene is a highly resilient synthetic rubber. Due to its outstanding resilience, it can be used for the manufacturing of golf balls. Heat buildup is lower in polybutadiene rubber based products subjected to repeated flexing during service. This property leads to its use in the sidewalls of car and truck tires. Low rolling resistance of this rubber also leads to its use in the tread portion of tires; however, relatively poor cold traction and cut growth resistance limits the use to blends with other types of rubber. Polybutadiene may be blended with, and crosslinked with, other types of rubber such as natural rubber styrene-butadiene rubber to optimize properties for tire treads and other applications. Polybutadiene rubber can be used in water seals for dams due to its low water absorption properties. Rubber bullets and road binders can be also produced by polybutadiene rubber. Polybutadiene is however, sensitive to oxidation and ozone owing to the reactivity of the double bond present in every repeat unit. Anti-oxidants are normally added to protect against cracking and deterioration.

### Copolymers

1,3-butadiene is normally copolymerized with other types of monomers such as styrene and acrylonitrile to form rubbers or plastics with various qualities. The most common form is styrene-butadiene copolymer, which is a commodity material for car tires. It is also used in block copolymers and tough thermoplastics such as ABS plastic. This way a copolymer material can be made with good stiffness, hardness, and toughness. Because the chains have a double bond in each and every repeat unit, the material is sensitive to ozone cracking.

### Processing

Polybutadiene rubber is seldom used alone, but is instead mixed with other rubbers. Polybutadiene is difficult to band in a two roll mixing mill. Instead, a thin sheet of polybutadiene may be prepared and kept separate. Then, after proper mastication of natural rubber, the polybutadiene rubber may be added to the two roll mixing mill. A similar practice may be adopted, for example, if polybutadiene is to be mixed with Styrene Butadiene Rubber (SBR). *Polybutadiene rubber may be added with Styrene as an impact modifier. High dosages may affect clarity of Styrene. In an internal mixer, natural rubber and/or styrene-butadiene rubber may be placed first, followed by polybutadiene. The plasticity of polybutadiene is not reduced by excessive mastication.

## Uses

### Tires

Polybutadiene is largely used in various parts of automobile tires; the manufacture of tires consumes about 70% of the world production of polybutadiene, with a majority of it being high *cis*. The polybutadiene is used primarily in the sidewall of truck tires, this helps to improve fatigue to failure life due to the continuous flexing during run. As a result, tires will not blow out in extreme service conditions. It is also used in the tread portion of giant truck tires to improve the abrasion, i.e. less wearing, and to run the tire comparatively cool, since the internal heat comes out quickly. Both parts are formed by extrusion.

Its main competitors in this application are styrene-butadiene rubber (SBR) and natural rubber. Polybutadiene has the advantage compared to SBR in its lower liquid-glass transition temperature, which gives it a high resistance to wear and a low rolling resistance. This gives the tires a long life and low fuel consumption. However, the lower transition temperature also lowers the friction on wet surfaces, which is why polybutadiene almost always is used in combination with any of the other two elastomers. About 1 kg of polybutadiene is used per tire in automobiles, and 3.3 kg in utility vehicles.

### Plastics

About 25% of the produced polybutadiene is used to improve the mechanical properties of plastics, in particular of high-impact polystyrene (HIPS) and to a lesser extent acrylonitrile butadiene styrene (ABS). The addition of between 4 and 12% polybutadiene to polystyrene transforms it from a fragile and delicate material to a ductile and resistant one.

The quality of the process is more important in the use in plastics than in tires, especially when it comes to colour and content of gels which have to be as low as possible. In addition, the products need to meet a list of health requirements due to its use in the food industry.

### *Golf Balls*

Most golf balls are made of an elastic core of polybutadiene surrounded by a layer of a harder material. Polybutadiene is preferred to other elastomers due to its high resilience.

The core of the balls are formed by compression molding with chemical reactions. First, polybutadiene is mixed with additives, then extruded, pressed using a calender and cut into pieces which are placed in a mold. The mold is subjected to high pressure and high temperature for about 30 minutes, enough time to vulcanize the material.

The golf ball production consumes about 20,000 tonnes of polybutadiene per year (1999).

### *Other Uses*

- Polybutadiene rubber may be used in the inner tube of hoses for sandblasting, along with natural rubber. The main idea is to increase resilience. This rubber can also be used in the cover of hoses, mainly pneumatic and water hoses.
- This rubber can also be used in railway pads, bridge blocks, etc.
- Polybutadiene rubber can be blended with nitrile rubber for easy processing. However large use may affect oil resistance of nitrile rubber.
- Polybutadiene is used in the manufacturing of the high-restitution toy Super Ball. Due to the high resilience property, 100% polybutadiene rubber based vulcanizate is used as crazy balls — i.e. a ball if dropped from 6th floor of a house will rebound up to 5½ to 6th floor (assuming no air resistance).
- It is also used as a fuel in combination with an oxidizer in various Solid Rocket Boosters such as Japan's H-IIB launch vehicle.

### *Production*

The annual production of polybutadiene was 2.0 million tons in 2003. This makes it the second most produced synthetic rubber by volume, behind the styrene-butadiene rubber (SBR).

The production processes of high *cis* polybutadiene and low *cis* used to be quite different and were carried out in separate plants. Lately, the trend has changed to use a single plant to produce as many different types of rubber as possible, including, low *cis* polybutadiene, high *cis* (with neodymium used as a catalyst) and SBR.

## Styrene-butadiene

Styrene-Butadiene or Styrene-Butadiene-Rubber (SBR) is a synthetic rubber copolymer consisting of styrene and butadiene. It has good abrasion resistance and good aging stability when protected by additives, and is widely used in car tires, where it may be blended with natural rubber. It was originally developed prior to World War II in Germany, but during the War was used extensively by the USA to replace natural rubber supplies from the far-east, that had been captured by the Japanese.

SBR can be produced by two basically different processes ; (a) from solution or (b) as emulsion. In the first instance the reaction is ionic polymerisation, in the emulsion polymerization case the reaction is via free radical polymerization.In that process; Low pressure reaction vessels are required and usually charged with styrene and butadiene, the two monomers, a free radical generator and a chain transfer agent such as an alkyl mercaptan and water. Mercaptans controls molecular weight and high viscosity product from forming. The anionic polymerization process is initiated by alkyl lithium and water not involved. High styrene content rubbers are harder but less rubbery.

It is not to be confused with a thermoplastic elastomer made from the same monomers, styrene-butadiene block copolymer.

### *Buna S*

The material was initially marketed with the brand name Buna S. It derives this name from: Bu for butadiene, Na for sodium (natrium in Latin, Natrium in German), and S for styrene.

### *Properties*

- Glass transition temperature approximately “55 °C (“67 °F), but varies with styrene content
- Possible temperature range of use: approximately -40 to 100 °C (-40 to 212 °F)
- Chemical Type: styrene-butadiene (copolymer)
- Trade names (common): GRS, Buna S, SBR
- Elongation (%): 450-500
- Shore A hardness: 50-90

- Major application characteristics: good physical properties; excellent abrasion resistance; but sensitive to oil, wastewater and ozone; electrical properties good, but not outstanding.

### *Applications*

The elastomer is used widely in pneumatic tires, shoe heels and soles, gaskets and even chewing gum. It is a commodity material which competes with natural rubber. Latex (emulsion) SBR is extensively used in coated papers, being one of the most cost-effective resins to bind pigmented coatings. It is also used in building applications, as a sealing and binding agent behind renders as an alternative to PVA, but is more expensive. In the latter application, it offers better durability, reduced shrinkage and increased flexibility, as well as being resistant to emulsification in damp conditions. SBR can be used to 'tank' damp rooms or surfaces, a process in which the rubber is painted onto the entire surface (sometimes both the walls, floor and ceiling) forming a continuous, seamless damp proof liner; a typical example would be a basement.

Additionally, it is used in some rubber cutting boards.

## Polyolefin

A polyolefin is a polymer produced from a simple olefin (also called an alkene with the general formula $C_nH_{2n}$) as a monomer. For example, polyethylene is the polyolefin produced by polymerizing the olefin ethylene. An equivalent term is polyalkene; this is a more modern term, although *polyolefin* is still used in the petrochemical industry. Polypropylene is another common polyolefin which is made from the olefin propylene.

### *Industrial Polyolefins*

- Thermoplastic polyolefins: polyethylene (PE), polypropylene (PP), polymethylpentene (PMP), polybutene-1 (PB-1);
- Polyolefin elastomers (POE): polyisobutylene (PIB), Ethylene propylene rubber (EPR), ethylene propylene diene Monomer (M-class) rubber (EPDM rubber)...

### *Properties*

Polyolefins are impossible to join by solvent cementing because they have excellent chemical resistance and can only be adhesively bonded after surface treatment because they have very low surface energies. They are also extremely inert chemically and exhibit decreased strength at lower temperatures.

A more specific type of olefin is a poly-*alpha*-olefin (or poly-α-olefin, sometimes abbreviated as PAO), a polymer made by polymerizing an *alpha*-olefin. An *alpha*-olefin (or α-olefin) is an alkene where the carbon-carbon double bond starts at the α-carbon atom, i.e. the double bond is between the #1 and #2 carbons in the molecule. Common alpha-olefins used as co-monomers to give a polymer alkyl branching groups are similar to 1-hexene or may be longer.

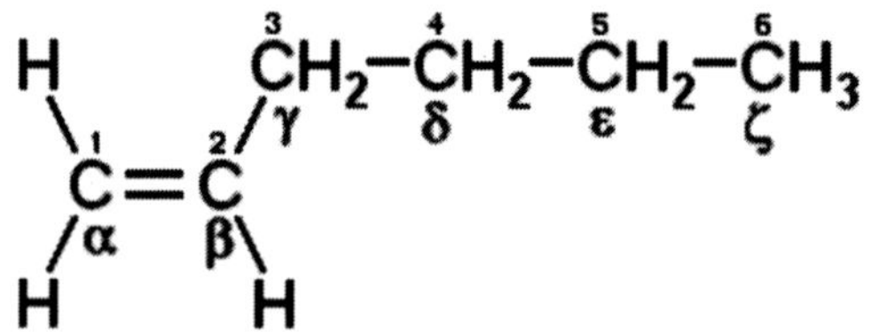

*1-hexene, an Example of an Alpha-olefin*

Many poly-alpha-olefins have flexible alkyl branching groups on every other carbon of their polymer backbone chain. These alkyl groups, which can shape themselves in numerous conformations, make it very difficult for the polymer molecules to line themselves up side-by-side in an orderly way. Therefore, many poly-alpha-olefins do not crystallize or solidify easily and are able to remain oily, viscous liquids even at lower temperatures. Low molecular weight poly-*alpha*-olefins are useful as synthetic lubricants such as synthetic motor oils for vehicles used in a wide temperature range. Even polyethylenes copolymerized with a small amount of alpha-olefins (such as 1-hexene, 1-octene, or longer) are more flexible than simple straight chain high density polyethylene, which has no branching. The methyl branch groups on a polypropylene polymer are not long enough to make typical commercial polypropylene more flexible than polyethylene.

### Uses

Polyolefin is used for blown film and heatshrink electrical insulation sleeves for crimped wire terminals, as well as rash guards or under garments for wetsuits. Polyolefin elastomer POE is used as a main ingredient in the molded flexible foam technology such as in the fabrication of self skinned footwear (for example, Crocs shoes), seat cushions, arm rests, spa pillows, etc. Hydrogenated PAO is used as a radar coolant. Head makes polyolefin tennis strings.

## Alpha-olefin

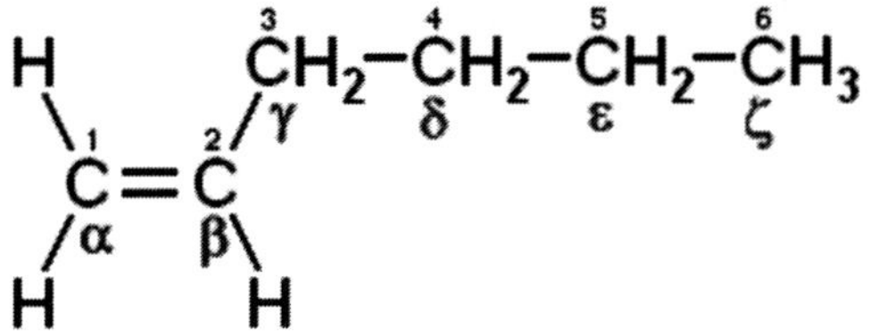

*Hex-1-ene, a Typical Alpha-olefin.*

The blue numbers show the IUPAC numbering of the atoms in the backbone chain of the molecule. The red symbols show the common nomenclature labeling of the main chain atoms. The double bond of an alpha olefin is between the #1 and #2 (IUPAC) or $\alpha$ and $\beta$ (common) carbon atoms. Alpha-olefins (or $\alpha$-olefins) are a family of organic compounds which are olefins or alkenes with a chemical formula $C_xH_{2x}$, distinguished by having a double bond at the primary or alpha ($\alpha$) position. This location of a double bond enhances the reactivity of the compound and makes it useful for a number of applications.

There are two types of alpha-olefins, branched and linear (or normal). The chemical properties of branched alpha-olefins with a branch at either the second (vinylidene) or the third carbon number are significantly different from the properties of linear alpha-olefins and those with branches on the fourth carbon number and further from the start of the chain.

## Benzene

Benzene is an organic chemical compound. It is composed of 6 carbon atoms in a ring, with 1 hydrogen atom attached to each carbon atom, with the molecular formula $C_6H_6$. Benzene is a natural constituent of crude oil, and is one of the most basic petrochemicals. Benzene is an aromatic hydrocarbon and the second [*n*]-annulene, a cyclic hydrocarbon with a continuous pi bond. It is sometimes abbreviated Ph–H. Benzene is a colorless and highly flammable liquid with a sweet smell. Because it is a known carcinogen, its use as an additive in gasoline is now limited, but it is an important industrial solvent and precursor to basic industrial chemicals including drugs, plastics, synthetic rubber, and dyes.

### *Structure*

Benzene represents a special problem in that, to account for all the bonds, there must be alternating double carbon bonds:

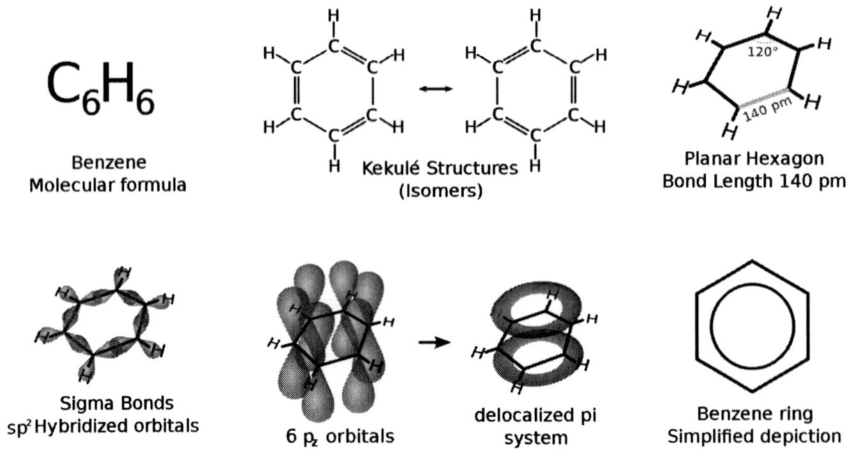


*The various representations of benzene*

X-ray diffraction shows that all of six carbon-carbon bonds in benzene are of the same length of 140 picometres (pm). The C–C bond lengths are greater than a double bond (135 pm) but shorter than a single bond (147 pm). This intermediate distance is consistent with electron delocalization: the electrons for C–C bonding are distributed equally between each of the six carbon atoms. The molecule is planar. One representation is that the structure exists as a superposition of so-called resonance structures, rather than either form individually. The delocalization of electrons is one explanation for the thermodynamic stability of benzene and related aromatic compounds. It is likely that this stability contributes to the peculiar molecular and chemical properties known as aromaticity. To indicate the delocalized nature of the bonding, benzene is often depicted with a circle inside a hexagonal arrangement of carbon atoms:

The delocalized picture of benzene has been contested by Cooper, Gerratt and Raimondi in their article published in 1986 in the journal Nature. They showed that the electrons in benzene are almost certainly localized, and the aromatic properties of benzene originate from spin coupling rather than electron delocalization. This view has been supported in the next-year Nature issue, but it has been slow to permeate the general chemistry community.

As is common in organic chemistry, the carbon atoms in the diagram above have been left unlabeled. Realizing each carbon has 2p electrons, each carbon donates an electron into the delocalized ring above and below the benzene ring. It is the side-on overlap of p-orbitals that produces the pi clouds.

Derivatives of benzene occur sufficiently often as a component of organic molecules that there is a Unicode symbol in the Miscellaneous Technical block with the code U+232C (,#) to represent it with three double bonds, and U+23E3 (ã#) for a delocalized version.

### Substituted Benzene Derivatives

Many important chemical compounds are derived from benzene by replacing one or more of its hydrogen atoms with another functional group. Examples of simple benzene derivatives are phenol, toluene, and aniline, abbreviated PhOH, PhMe, and $PhNH_2$, respectively. Linking benzene rings gives biphenyl, $C_6H_5$–$C_6H_5$. Further loss of hydrogen gives "fused" aromatic hydrocarbons, such as naphthalene and anthracene. The limit of the fusion process is the hydrogen-free allotrope of carbon, graphite.

In heterocycles, carbon atoms in the benzene ring are replaced with other elements. The most important derivatives are the rings containing nitrogen. Replacing one CH with N gives the compound pyridine, $C_5H_5N$. Although benzene and pyridine are *structurally* related, benzene cannot be converted into pyridine. Replacement of a second CH bond with N gives, depending on the location of the second N, pyridazine, pyrimidine, and pyrazine.

### Production

Four chemical processes contribute to industrial benzene production: catalytic reforming, toluene hydrodealkylation, toluene disproportionation, and steam cracking. According to the ATSDR Toxicological Profile for benzene, between 1978 and 1981, catalytic reformats accounted for approximately 44–50% of the total U.S benzene production.

Until World War II, most benzene was produced as a by-product of coke production (or "coke-oven light oil") in the steel industry. However, in the 1950s, increased demand for benzene, especially from the growing polymers industry, necessitated the production of benzene from petroleum. Today, most benzene comes from the petrochemical industry, with only a small fraction being produced from coal.

### Catalytic Reforming

In catalytic reforming, a mixture of hydrocarbons with boiling points between 60–200 °C is blended with hydrogen gas and then exposed to a bifunctional platinum chloride or rhenium chloride catalyst at 500–525 °C and pressures ranging from 8–50 atm. Under these conditions, aliphatic hydrocarbons form rings and lose hydrogen to become aromatic hydrocarbons.

The aromatic products of the reaction are then separated from the reaction mixture (or reformate) by extraction with any one of a number of solvents, including diethylene glycol or sulfolane, and benzene is then separated from the other aromatics by distillation. The extraction step of aromatics from the reformate is designed to produce aromatics with lowest non-aromatic components. So-called BTX (benzene-toluene-xylene) process consists of such extraction and distillation steps. One such widely used process from UOP was licensed to producers and called the Udex process.

In similar fashion to this catalytic reforming, UOP and BP commercialized a method from LPG (mainly propane and butane) to aromatics.

### Toluene Hydrodealkylation

Toluene hydrodealkylation converts toluene to benzene. In this hydrogen-intensive process, toluene is mixed with hydrogen, then passed over a chromium, molybdenum, or platinum oxide catalyst at 500–600 °C and 40–60 atm pressure. Sometimes, higher temperatures are used instead of a catalyst (at the similar reaction condition). Under these conditions, toluene undergoes dealkylation to benzene and methane:

$$C_6H_5CH_3 + H_2 \rightarrow C_6H_6 + CH_4$$

This irreversible reaction is accompanied by an equilibrium side reaction that produces biphenyl (aka diphenyl) at higher temperature:

$$2\ C_6H_6 \rightleftharpoons H_2 + C_6H_5\text{–}C_6H_5$$

If the raw material stream contains much non-aromatic components (paraffins or naphthenes), those are likely decomposed to lower hydrocarbons such as methane, which increases the consumption of hydrogen.

A typical reaction yield exceeds 95%. Sometimes, xylenes and heavier aromatics are used in place of toluene, with similar efficiency.

This is often called "on-purpose" methodology to produce benzene, compared to conventional BTX (benzene-toluene-xylene) processes.

### *Toluene Disproportionation*

Where a chemical complex has similar demands for both benzene and xylene, then toluene disproportionation (TDP) may be an attractive alternative to the toluene hydrodealkylation. In the broad sense, 2 toluene molecules are reacted and the methyl groups rearranged from one toluene molecule to the other, yielding one benzene molecule and one xylene molecule.

Given that demand for *para*-xylene (*p*-xylene) substantially exceeds demand for other xylene isomers, a refinement of the TDP process called Selective TDP (STDP) may be used. In this process, the xylene stream exiting the TDP unit is approximately 90% paraxylene. In some current catalytic systems, even the benzene-to-xylenes ratio is decreased (more xylenes) when the demand of xylenes is higher.

### *Steam Cracking*

Steam cracking is the process for producing ethylene and other alkenes from aliphatic hydrocarbons. Depending on the feedstock used to produce the olefins, steam cracking can produce a benzene-rich liquid by-product called *pyrolysis gasoline*. Pyrolysis gasoline can be blended with other hydrocarbons as a gasoline additive, or distilled (in BTX process) to separate it into its components, including benzene.

### *Other Sources*

Trace amounts of benzene may result whenever carbon-rich materials undergo incomplete combustion. It is produced in volcanoes and forest fires, and is also a component of cigarette smoke. Benzene is a principal product from the combustion of PVC (polyvinyl chloride).

### *Uses—Early Uses*

In the 19th and early-20th centuries, benzene was used as an after-shave lotion because of its pleasant smell. Prior to the 1920s, benzene was frequently used as an industrial solvent, especially for degreasing metal. As its toxicity became obvious, benzene was supplanted by other solvents, especially toluene (methyl benzene), which has similar physical properties but is not as carcinogenic.

In 1903, Ludwig Roselius popularized the use of benzene to decaffeinate coffee. This discovery led to the production of Sanka (the letters "ka" in the brand name stand for *kaffein*). This process was later discontinued. Benzene was historically used as a significant component in many consumer products such as Liquid Wrench, several paint strippers, rubber cements, spot removers and other hydrocarbon-containing products. Some ceased manufacture of their benzene-containing formulations in about 1950, while others continued to use benzene as a component or significant contaminant until the late 1970s when leukemia deaths were found associated with Goodyear's Pliofilm production operations in Ohio. Until the late 1970s, many hardware stores, paint stores, and other retail outlets sold benzene in small cans, such as quart size, for general-purpose use. Many students were exposed to benzene in school and university courses while performing laboratory experiments with little or no ventilation in many cases. This very dangerous practice has been almost totally eliminated.

As a gasoline (petrol) additive, benzene increases the octane rating and reduces knocking. As a consequence, gasoline often contained several percent benzene before the 1950s, when tetraethyl lead replaced it as the most widely-used antiknock additive. With the global phaseout of leaded gasoline, benzene has made a comeback as a gasoline additive in some nations. In the United States, concern over its negative health effects and the possibility of benzene's entering the groundwater have led to stringent regulation of gasoline's benzene content, with limits typically around 1%. European petrol specifications now contain the same 1% limit on benzene content. The United States Environmental Protection Agencyý has new regulations that will lower the benzene content in gasoline to 0.62% in 2011.

### Current Uses

Today, benzene is used mainly as an intermediate to make other chemicals. Its most widely-produced derivatives include styrene, which is used to make polymers and plastics, phenol for resins and adhesives (via cumene), and cyclohexane, which is used in the manufacture of Nylon. Smaller amounts of benzene are used to make some types of rubbers, lubricants, dyes, detergents, drugs, explosives, napalm, and pesticides.

In both the US and Europe, 50% of benzene is used in the production of ethylbenzene/styrene, 20% is used in the production of cumene, and about 15% of benzene is used in the production of cyclohexane (eventually to nylon).

In laboratory research, toluene is now often used as a substitute for benzene. The solvent-properties of the two are similar, but toluene is less toxic and has a wider liquid range.

Benzene has been used as a basic research tool in a variety of experiments including analysis of a two-dimensional gas.

## Reactions

The most common reactions that benzene undergoes are substitution reactions.

- Electrophilic aromatic substitution is a general method of derivatizing benzene. Benzene is sufficiently nucleophilic that it undergoes substitution by acylium ions or alkyl carbocations to give substituted derivatives.
- The Friedel-Crafts alkylation involves the alkylation of benzene (and many other aromatic rings) using an alkyl halide in the presence of a strong Lewis acid catalyst.
- The Friedel-Crafts acylation is a specific example of electrophilic aromatic substitution. The reaction involves the acylation of benzene (or many other aromatic rings) with an acyl chloride using a strong Lewis acid catalyst such as aluminium chloride or Iron(III) chloride, which act as a halogen carrier.
- Sulfonation. The most common method involves mixing sulfuric acid with sulfate, a mixture called fuming sulfuric acid. The sulfuric acid protonates the sulfate, giving the sulfur atom a permanent, rather than resonance stabilized positive formal charge. This molecule is very electrophillic and Electrophillic Aromatic Substitution then occurs.
- Nitration: Benzene undergoes nitration with nitronium ions ($NO_2^+$) as the electrophile. Thus, warming benzene at 50–55 °C, with a combination of concentrated sulfuric and nitric acid to produce the electrophile, gives nitrobenzene.
- Hydrogenation (reduction): Benzene and derivatives convert to cyclohexane and derivatives when treated with hydrogen at 450 K and 10 atm of pressure with a finely divided nickel catalyst.
- Benzene is an excellent ligand in the organometallic chemistry of low-valent metals. Important examples include the sandwich and half-sandwich complexes, respectively, $Cr(C_6H_6)_2$ and $[RuCl_2(C_6H_6)]_2$.

## Environmental Transformation

Even if it is not a common substrate for the metabolism of

organisms, benzene could be oxidized by both bacteria and eukaryotes. In bacteria, dioxygenase enzyme can add an oxygen molecule to the ring, and the unstable product is immediately reduced (by NADH) to a cyclic diol with two double bonds, breaking the aromaticity. Next, the diol is newly reduced by NADH to catechol.

The catechol is then metabolized to acetyl CoA and succinyl CoA, used by organisms mainly in the Krebs Cycle for energy production.

### Health Effects

Benzene causes cancer and other illnesses. Benzene is a "notorious cause" of bone marrow failure. "Vast quantities of epidemiologic, clinical, and laboratory data" link benzene to aplastic anemia, acute leukemia, and bone marrow abnormalities.

The American Petroleum Institute (API) stated in 1948 that "it is generally considered that the only absolutely safe concentration for benzene is zero." The US Department of Health and Human Services (DHHS) classifies benzene as a human carcinogen. Long-term exposure to excessive levels of benzene in the air causes leukemia, a potentially fatal cancer of the blood-forming organs, in susceptible individuals. In particular, Acute myeloid leukemia or acute non-lymphocytic leukaemia (AML & ANLL) is not disputed to be caused by benzene. IARC rated benzene as "known to be carcinogenic to humans" (Group 1).

Outdoor air may contain low levels of benzene from automobile service stations, wood smoke, tobacco smoke, the transfer of gasoline, exhaust from motor vehicles, and industrial emissions. About 50% of the entire nationwide (United States) exposure to benzene results from smoking tobacco or from exposure to tobacco smoke.

Vapours from products that contain benzene, such as glues, paints, furniture wax, and detergents, can also be a source of exposure, although many of these have been modified or reformulated since the late 1970s to eliminate or reduce the benzene content. Air around hazardous waste sites or gas stations may contain higher levels of benzene. Because petroleum hydrocarbon products are complex mixtures of chemicals, risk assessments for these products, in general, focus on specific toxic constituents. The petroleum constituents of primary interest to human health have been the aromatic hydrocarbons (i.e., benzene, ethylbenzene, toluene, and xylenes). OSHA requires that a mixture "shall be assumed to present a carcinogenic hazard if it contains a component in concentrations of 0.1% or greater, which is considered to be a carcinogen.

The short-term breathing of high levels of benzene can result in death; low levels can cause drowsiness, dizziness, rapid heart rate, headaches, tremors, confusion, and unconsciousness. Eating or drinking foods containing high levels of benzene can cause vomiting, irritation of the stomach, dizziness, sleepiness, convulsions, and death.

The major effects of benzene are manifested via chronic (long-term) exposure through the blood. Benzene damages the bone marrow and can cause a decrease in red blood cells, leading to anemia. It can also cause excessive bleeding and depress the immune system, increasing the chance of infection. Benzene causes leukemia and is associated with other blood cancers and pre-cancers of the blood.

Human exposure to benzene is a global health problem. Benzene targets liver, kidney, lung, heart and the brain and can cause DNA strand breaks, chromosomal damage, etc. Benzene causes cancer in both animals and humans. Benzene was first reported to induce cancer in humans in the 1920s. The chemical industry claims it was not until 1979 that the cancer-inducing properties were determined "conclusively" in humans, despite many references to this fact in the medical literature. Industry exploited this "discrepancy" and tried to discredit animal studies that showed that benzene causes cancer, saying that they are not relevant to humans. Benzene has been shown to cause cancer in both sexes of multiple species of laboratory animals exposed via various routes.

Some women having breathed high levels of benzene for many months had irregular menstrual periods and a decrease in the size of their ovaries. Benzene exposure has been linked directly to the neural birth defects spina bifida and anencephaly. Men exposed to high levels of benzene are more likely to have an abnormal amount of chromosomes in their sperm, which impacts fertility and fetal development.

Animal studies have shown low birth weights, delayed bone formation, and bone marrow damage when pregnant animals breathed benzene.

Benzene has been connected to a rare form of kidney cancer in two separate studies, one involving tank truck drivers, and the other involving seamen on tanker vessels, both carrying benzene-laden chemicals.

### Exposure to Benzene

Workers in various industries that make or use benzene may be at risk for being exposed to high levels of this carcinogenic chemical. Industries that involve the use of benzene include the rubber industry, oil refineries, coke and chemical plants, shoe manufacturers, and gasoline-related industries. Downstream petroleum industry operations include the following categories: refinery, pipeline, marine, rail, bulk terminals and trucks, service stations, underground storage tanks, tank cleaning, and site characterization and remediation.

Exposure of the general population to benzene mainly occurs through breathing, the major sources of benzene being tobacco smoke (about 50%) as well as automobile service stations, exhaust from motor vehicles and industrial emissions (about 20% altogether). Vapours (or gases) from products that contain benzene, such as glues, paints, furniture wax, and detergents, can also be a source of exposure. The average smoker (32 cigarettes per day) takes in about 1.8 milligrams (mg) of benzene per day. This amount is about 10 times the average daily intake of benzene by nonsmokers.

In 1987, OSHA estimated that about 237,000 workers in the United States were potentially exposed to benzene, but it is not known if this number has substantially changed since then.

Water and soil contamination are important pathways of concern for transmission of benzene contact. In the US alone, there are approximately 100,000 different sites that have benzene soil or groundwater contamination. In 2005, the water supply to the city of Harbin in China with a population of almost nine million people, was cut off because of a major benzene exposure. Benzene leaked into the Songhua River, which supplies drinking water to the city, after an explosion at a China National Petroleum Corporation (CNPC) factory in the city of Jilin on 13 November.

In March 2006, the official Food Standards Agency in Britain conducted a survey of 150 brands of soft drinks. It found that four contained benzene levels above World Health Organization limits. The affected batches were removed from sale.

### Benzene Exposure Limits

The United States Environmental Protection Agency has set a maximum contaminant level (MCL) for benzene in drinking water at 0.005 mg/L (5 ppb), as promulgated via the National Primary Drinking

Water Regulations. This regulation is based on preventing benzene leukemogenesis. The maximum contaminant level goal (MCLG), a nonenforceable health goal that would allow an adequate margin of safety for the prevention of adverse effects, is zero benzene concentration in drinking water. The EPA requires that spills or accidental releases into the environment of 10 pounds (4.5 kg) or more of benzene be reported to the EPA.

The US Occupational Safety and Health Administration (OSHA) has set a permissible exposure limit of 1 part of benzene per million parts of air (1 ppm) in the workplace during an 8-hour workday, 40-hour workweek. The short term exposure limit for airborne benzene is 5 ppm for 15 minutes. These legal limits were based on studies demonstrating compelling evidence of health risk to workers exposed to benzene. The risk from exposure to 1 ppm for a working lifetime has been estimated as 5 excess leukemia deaths per 1,000 employees exposed. (This estimate assumes no threshold for benzene's carcinogenic effects.) OSHA has also established an action level of 0.5 ppm to encourage even lower exposures in the workplace.

The National Institute for Occupational Safety and Health (NIOSH) revised the Immediately Dangerous to Life or Health (IDLH) concentration for benzene to 500 ppm. The current NIOSH definition for an IDLH condition, as given in the NIOSH Respirator Selection Logic, is one that poses a threat of exposure to airborne contaminants when that exposure is likely to cause death or immediate or delayed permanent adverse health effects or prevent escape from such an environment [NIOSH 2004]. The purpose of establishing an IDLH value is (1) to ensure that the worker can escape from a given contaminated environment in the event of failure of the respiratory protection equipment and (2) is considered a maximum level above which only a highly reliable breathing apparatus providing maximum worker protection is permitted [NIOSH 2004]. In September 1995, NIOSH issued a new policy for developing recommended exposure limits (RELs) for substances, including carcinogens. Because benzene can cause cancer, NIOSH recommends that all workers wear special breathing equipment when they are likely to be exposed to benzene at levels exceeding the REL (10-hour) of 0.1 ppm. The NIOSH STEL (15 min) is 1 ppm.

American Conference of Governmental Industrial Hygienists (ACGIH) adopted Threshold Limit Values (TLVs) for benzene at 0.5 ppm TWA and 2.5 ppm STEL.

Under New Jersey's Right-to-Know law, respiratory protection for benzene is discussed. As stated, improper use of respirators is dangerous. Respirators should only be used when there is a written respiratory program in place as described in the OSHA Respiratory Protection Standard (29 CFR 1910.134). The employer is to develop and implement a written respiratory protection program with required worksite-specific procedures and elements for required respirator use. This program must be administered by a suitably trained program administrator. For benzene:

- If there is a potential for exposure to 0.1 ppm, a NIOSH-approved half face respirator must be worn with an organic vapour cartridge (APF 10).
- If there is a potential for exposure to 0.5 ppm, a NIOSH-approved full face respirator must be worn with an organic vapour cartridge (APF 50).
- Where the potential exists for exposure to over 5 ppm, a NIOSH-approved air-supplied respirator with a full facepiece operated under pressure-demand or other positive-pressure mode must be used (APF 1,000).

### Exposure Monitoring

Airborne exposure monitoring for benzene must be conducted in order to properly assess personal exposures and effectiveness of engineering controls. Initial exposure monitoring should be conducted by an industrial hygienist or person specifically trained and experienced in sampling techniques. Contact an AIHA Accredited Laboratory for advice on sampling methods.

Each employer with a place of employment where occupational exposures to benzene occur shall monitor each of these workplaces and work operations to determine accurately the airborne concentrations of benzene to which employees may be exposed. Representative 8-hour TWA employee exposures need to be determined on the basis of one sample or samples representing the full shift exposure for each job classification in each work area. Unless air samples are taken frequently, the employer does not know the concentration and would not know how much of a protection factor is needed.

In providing consultation on work safety during oil clean-up operations following the Deepwater Horizon accident, OSHA has worked with a number of other government agencies to protect Gulf cleanup workers. OSHA partnered with the NIOSH to issue "Interim Guidance for Protecting Deepwater Horizon Response Workers and Volunteers" and recommend measures that should be taken to protect workers from a variety of different health hazards that these workers face. OSHA conceded that it recognizes that most of its PELs are outdated and inadequate measures of worker safety. In characterizing worker exposure, OSHA instead relies on more up-to-date recommended protective limits set by organizations such as NIOSH, the ACGIH, and the American Industrial Hygiene Association (AIHA), and not on the older, less protective PELS. Results of air monitoring are compared to the lowest known Occupational Exposure Limit for the listed contaminant for purposes of risk assessment and protective equipment recommendations.

### Biomarkers of Exposure

Several tests can determine exposure to benzene. Benzene itself can be measured in breath, blood or urine, but such testing is usually limited to the first 24 hours post-exposure due to the relatively rapid removal of the chemical by exhalation or biotransformation. Most persons in developed countries have measureable baseline levels of benzene and other aromatic petroleum hydrocarbons in their blood. In the body, benzene is enzymatically converted to a series of oxidation products including muconic acid, phenylmercapturic acid, phenol, catechol, hydroquinone and 1,2,4-trihydroxybenzene. Most of these metabolites have some value as biomarkers of human exposure, since they accumulate in the urine in proportion to the extent and duration of exposure, and they may still be present for some days after exposure has ceased. The current ACGIH biological exposure limits for occupational exposure are 500 μg/g creatinine for muconic acid and 25 μg/g creatinine for phenylmercapturic acid in an end-of-shift urine specimen.

### Methods of Excretion

Most inhaled benzene is not metabolized. Inhaled benzene is primarily expelled unchanged through exhalation. In a human study 16.4 to 41.6% of retained benzene was eliminated through the lungs within five to seven hours after a two- to three-hour exposure to 47 to 110 ppm and only 0.07 to 0.2% of the remaining benzene was excreted unchanged in the urine. After exposure to 63 to 405 mg/m3 of benzene for 1 to 5 hours, 51 to 87% was excreted in the urine as phenol over a period of 23 to 50 hours. In another human study, 30% of absorbed dermally applied benzene, which is primarily metabolized in the liver, was excreted as phenol in the urine.

### *Molecular Toxicology*

The paradigm of toxicological assessment of benzene is slowly shifting towards the domain of molecular toxicology as it allows understanding of fundamental biological mechanisms in a better way. Glutathione seems to play an important role by protecting against benzene-induced DNA breaks and it is being identified as a new biomarker for exposure and effect. Benzene causes chromosomal aberrations in the peripheral blood leukocytes and bone marrow explaining the higher incidence of leukemia and multiple myeloma caused by chronic exposure. These aberrations can be monitored using fluorescent in situ hybridization (FISH) with DNA probes to assess the effects of benzene along with the hematological tests as markers of hematotoxicity. Benzene metabolism involves enzymes coded for by polymorphic genes. Studies have shown that genotype at these loci may influence susceptibility to the toxic effects of benzene exposure. Individuals carrying variant of NAD(P)H:quinone oxidoreductase 1 (NQO1), microsomal epoxide hydrolase (EPHX) and deletion of the glutathione S-transferase T1 (GSTT1) showed a greater frequency of DNA single-stranded breaks.

### *Biological Oxidation and Carcinogenic Activity*

One way of understanding the carcinogenic effects of benzene is by examining the products of biological oxidation. Pure benzene, for example, oxidizes in the body to produce an epoxide, benzene oxide, which is not excreted readily and can interact with DNA to produce harmful mutations.

According to the Agency for Toxic Substances and Disease Registry (ATSDR) (2007), benzene is both an anthropogenically produced and naturally occurring chemical from processes that include: volcanic eruptions, wild fires, synthesis of chemicals such as phenol, production of synthetic fibres, and fabrication of rubbers, lubricants, pesticides, medications, and dyes. The major sources of benzene exposure are tobacco smoke, automobile service stations, exhaust from motor vehicles, and industrial emissions; however, ingestion and dermal absorption of benzene can also occur through contact with contaminated water. Benzene is hepatically metabolized and excreted in the urine. Measurement of air and water levels of benzene is accomplished through collection via activated charcoal tubes, which are then analyzed with a gas chromatograph. The measurement of benzene in humans can be accomplished via urine, blood, and breath tests; however, all of these have their limitations because benzene is rapidly metabolized in the human body into by-products called metabolites.

OSHA regulates levels of benzene in the workplace. The maximum allowable amount of benzene in workroom air during an 8-hour workday, 40-hour workweek is 1 ppm. Because benzene can cause cancer, NIOSH recommends that all workers wear special breathing equipment when they are likely to be exposed to benzene at levels exceeding the recommended (8-hour) exposure limit of 0.1 ppm.

## History—Discovery

The word "benzene" derives historically from "gum benzoin", sometimes called "benjamin" (i.e., benzoin resin), an aromatic resin known to European pharmacists and perfumers since the 15th century as a product of southeast Asia. "Benzoin" is itself a corruption of the Arabic expression *"luban jawi"*, or "frankincense of Java". An acidic material was derived from benzoin by sublimation, and named "flowers of benzoin", or benzoic acid. The hydrocarbon derived from benzoic acid thus acquired the name benzin, benzol, or benzene.

Michael Faraday first isolated and identified benzene in 1825 from the oily residue derived from the production of illuminating gas, giving it the name *bicarburet of hydrogen.*

In 1833, Eilhard Mitscherlich produced it via the distillation of benzoic acid (from gum benzoin) and lime. He gave the compound the name *benzin.*

In 1836, the French chemist Auguste Laurent named the substance "phène"; this is the root of the word phenol, which is hydroxylated benzene, and phenyl, which is the radical formed by abstraction of a hydrogen atom (free radical H*) from benzene.

In 1845, Charles Mansfield, working under August Wilhelm von Hofmann, isolated benzene from coal tar. Four years later, Mansfield began the first industrial-scale production of benzene, based on the coal-tar method. Gradually the sense developed among chemists that substances related to benzene represent a diverse chemical family. In 1855 August Wilhelm Hofmann used the word "aromatic" to designate this family relationship, after a characteristic property of many of its members.

## *Ring Formula*

The empirical formula for benzene was long known, but its highly polyunsaturated structure, with just one hydrogen atom for each carbon atom, was challenging to determine. Archibald Scott Couper in 1858 and Joseph Loschmidt in 1861 suggested possible structures that contained multiple double bonds or multiple rings, but too little evidence was then available to help chemists decide on any particular structure.

In 1865, the German chemist Friedrich August Kekulé published a paper in French (for he was then teaching in Francophone Belgium) suggesting that the structure contained a six-membered ring of carbon atoms with alternating single and double bonds. The next year he published a much longer paper in German on the same subject. Kekulé used evidence that had accumulated in the intervening years—namely, that there always appeared to be only one isomer of any monoderivative of benzene, and that there always appeared to be exactly three isomers of every diderivative—now understood to correspond to the ortho, meta, and para patterns of arene substitution—to argue in support of his proposed structure. Kekulé's symmetrical ring could explain these curious facts, as well as benzene's 1:1 carbon-hydrogen ratio.

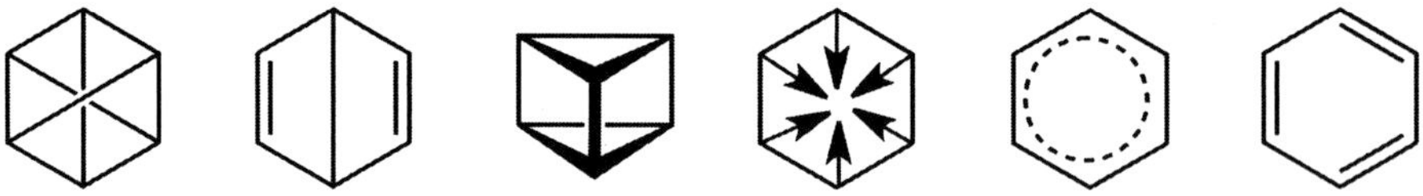

Historic benzene formulae (from left to right) by Claus (1867), Dewar (1867), Ladenburg (1869), Armstrong (1887), Thiele (1899) and Kekulé (1865). Dewar benzene and prismane are different chemicals that have Dewar's and Ladenburg's structures. Thiele and Kekulé's structures are used today.

The new understanding of benzene, and hence of all aromatic compounds, proved to be so important for both pure and applied chemistry that in 1890 the German Chemical Society organized an elaborate appreciation in Kekulé's honor, celebrating the twenty-fifth anniversary of his first benzene paper. Here Kekulé spoke of the creation of the theory. He said that he had discovered the ring shape of the benzene molecule after having a reverie or day-dream of a snake seizing its own tail (this is a common symbol in many ancient cultures known as the Ouroboros or Endless knot). This vision, he said, came to him after years of studying the nature of carbon-carbon bonds. This was 7 years after he had solved the problem of how carbon atoms could bond to up to four other atoms at the same time. It is curious that a similar, humorous depiction of benzene had appeared in 1886 in the *Berichte der Durstigen Chemischen Gesellschaft* (Journal of the Thirsty Chemical Society), a parody of the *Berichte der Deutschen Chemischen Gesellschaft*, only the parody had monkeys seizing each other in a circle, rather than snakes as in Kekulé's anecdote. Some historians have suggested that the parody was a lampoon of the snake anecdote, possibly already well-known through oral transmission even if it had not yet appeared in print. Others have speculated that Kekulé's story in 1890 was a re-parody of the monkey spoof, and was a mere invention rather than a recollection of an event in his life. Kekulé's 1890 speech in which these anecdotes appeared has been translated into English. If one takes the anecdote as the memory of a real event, circumstances mentioned in the story suggest that it must have happened early in 1862.

The cyclic nature of benzene was finally confirmed by the crystallographer Kathleen Lonsdale in 1929.

# 2

# Petrochemicals of Petroleum

## Styrene

Styrene, also known as vinyl benzene, is an organic compound with the chemical formula $C_6H_5CH=CH_2$. This derivative of benzene is a colorless oily liquid that evaporates easily and has a sweet smell, although high concentrations confer a less pleasant odor. Styrene is the precursor to polystyrene and several copolymers. Approximately 15 billion pounds are produced annually. On 10 June 2011, the US National Toxicology Program has described styrene as "reasonably anticipated to be a human carcinogen". However, an academic panel recently reviewed the relevant scientific literature and concluded that "The available epidemiologic evidence does not support a causal relationship between styrene exposure and any type of human cancer."

### *Occurrence, History, and Use*

Styrene is named for "styrax", the resin from a Turkish tree, the Oriental sweetgum *(Liquidambar orientalis)*, from which it was first isolated, and not for the tropical *Styrax* trees from which benzoin resin is produced. Low levels of styrene occur naturally in many kinds of plants, as well as a variety of foods such as fruits, vegetables, nuts, beverages, and meats.

The production of styrene in the United States increased dramatically during the 1940s, when it was popularized as a feedstock for synthetic rubber.

The presence of the vinyl group allows styrene to polymerize. Commercially significant products include polystyrene, ABS, styrene-butadiene (SBR) rubber, styrene-butadiene latex, SIS (styrene-isoprene-styrene), S-EB-S (styrene-ethylene/butylene-styrene), styrene-divinylbenzene (S-DVB), styrene-acrylonitrile resin (SAN) and unsaturated polyesters. These materials are used in rubber, plastic, insulation, fiberglass, pipes, automobile and boat parts, food containers, and carpet backing.

### Production

Styrene is produced in industrial quantities from ethylbenzene, which is in turn prepared on a large scale by alkylation of benzene with ethylene.

### Dehydrogenation of Ethylbenzene

Styrene is most commonly produced by the catalytic dehydrogenation of ethylbenzene. Ethylbenzene is mixed in the gas phase with 10–15 times its volume in high-temperature steam, and passed over a solid catalyst bed. Most ethylbenzene dehydrogenation catalysts are based on iron(III) oxide, promoted by several percent potassium oxide or potassium carbonate.

Steam serves several roles in this reaction. It is the source of heat for powering the endothermic reaction, and it removes coke that tends to form on the iron oxide catalyst through the water gas shift reaction. The potassium promoter enhances this decoking reaction. The steam also dilutes the reactant and products, shifting the position of chemical equilibrium towards products. A typical styrene plant consists of two or three reactors in series, which operate under vacuum to enhance the conversion and selectivity. Typical per-pass conversions are ca. 65% for two reactors and 70-75% for three reactors. Selectivity to styrene is 93-97%. The main byproducts are benzene and toluene. Because styrene and ethylbenzene have similar boiling points (145 and 136 °C, respectively), their separation requires tall distillation towers and high return/reflux ratios. At its distillation temperatures, styrene tends to polymerize. To minimize this problem, early styrene plants added elemental sulfur to inhibit the polymerization. During the 1970s, new free radical inhibitors consisting of nitrated phenol-based retarders were developed. More recently, a number of additives have been developed that exhibit superior inhibition against polymerization. However, the nitrated phenols are still widely used because of their relatively low cost. These reagents are added prior to the distillation.

Improving conversion and so reducing the amount of ethylbenzene that must be separated is the chief impetus for researching alternative routes to styrene. Other than the POSM process, none of these routes like obtaining styrene from butadiene have been commercially demonstrated.

### *Via Ethylbenzenehydroperoxide*

Commercially styrene is also co-produced with propylene oxide in a process known as POSM (Lyondell Chemical Company) or SM/PO (Shell) for styrene monomer/propylene oxide. In this process ethylbenzene is treated with oxygen to form the ethylbenzene hydroperoxide. This hydroperoxide is then used to oxidize propylene to propylene oxide. The resulting 2-phenylethanol is dehydrated to give styrene:

$$C_6H_5CH_2CH_3 + O_2 \rightarrow C_6H_5CH_2CH_2O_2H$$

$$C_6H_5CH_2CH_2O_2H + CH_3CH{=}CH_2 \rightarrow$$

$$C_6H_5CH_2CH_2OH + CH_3CHCH_2O$$

$$C_6H_5CH_2CH_2OH \rightarrow C_6H_5CH{=}CH_2 + H_2O$$

### *Laboratory Synthesis*

A laboratory synthesis of styrene entails the decarboxylation of cinnamic acid. Styrene was first prepared by this method.

### *Other Methods*

Styrene can be produced from toluene and methanol, which are cheaper raw materials than those in the conventional process. Historically, however, this process has suffered from low selectivity due to competing decomposition of methanol. Exelus Inc. claims to have developed this process with commercially viable selectivities, at 400-425 °C and atmospheric pressure, by forcing these components through a proprietary zeolitic catalyst. It is reported that an approximately 9:1 mixture of styrene and ethylbenzene is obtained, with a total styrene yield of over 60%.

Another developing route to styrene is via benzene and ethane. This process is being developed by Snamprogetti S.p.A. and Dow. Ethane, along with ethylbenzene, is fed to a dehydrogenation reactor with a catalyst capable of simultaneously producing styrene and ethylene. The dehydrogenation effluent is cooled and separated and the ethylene stream is recycled to the alkylation unit. The process attempts to overcome previous shortcomings in earlier attempts to develop production of styrene from ethane and benzene, such as inefficient recovery of aromatics, production of high levels of heavies and tars, and inefficient separation of hydrogen and ethane. Development of the process is ongoing.

### Health Effects

Styrene is regarded as a "hazardous chemical", especially in case of eye contact, but also in case of skin contact, of ingestion and of inhalation, according to several sources. The US EPA has described styrene to be "a suspected toxin to the gastrointestinal tract, kidney, and respiratory system, among others." On 10 June 2011, the US National Toxicology Program has described styrene as "reasonably anticipated to be a human carcinogen". However, an academic panel, funded by the styrene industry, recently reviewed the relevant scientific literature and concluded that "The available epidemiologic evidence does not support a causal relationship between styrene exposure and any type of human cancer."

The U.S. EPA does not have a cancer classification for styrene, but currently is evaluating styrene's cancer-causing potential through its Integrated Risk Information System (IRIS) program. The U.S. National Toxicology Program of the U.S. Department of Health and Human Services also currently is evaluating styrene's potential toxicity To date, no regulatory body anywhere in the world has classified styrene as a known human carcinogen, although several refer to it in various contexts as a possible or potential human carcinogen. The International Agency for Research on Cancer considers styrene to be "possibly carcinogenic to humans.". Chronic exposure to styrene leads to tiredness/lethargy, memory deficits, headaches and vertigo.

According to the Styrene Information and Research Centre (an organization representing nearly all of the "North American styrene industry"), polystyrene plastic neither contains, nor breaks down into bisphenol A (BPA), a chemical used in plastic compounds that leads to developmental and reproductive problems in both adults and children.

## Polystyrene

Polystyrene also known as Thermocole, abbreviated following ISO Standard PS, is an aromatic polymer made from the monomer styrene, a liquid hydrocarbon that is manufactured from petroleum by the chemical industry. Polystyrene is one of the most widely used plastics, the scale being several billion kilograms per year.

Polystyrene is a thermoplastic substance, which is in solid (glassy) state at room temperature, but flows if heated above its glass transition temperature of about 100 °C (for molding or extrusion), and becomes solid again when cooled. Pure solid polystyrene is a colorless, hard plastic with limited flexibility. It can be cast into molds with fine detail. Polystyrene can be transparent or can be made to take on various colours.

Solid polystyrene is used, for example, in disposable cutlery, plastic models, CD and DVD cases, and smoke detector housings. Products made from foamed polystyrene are nearly ubiquitous, for example packing materials, insulation, and foam drink cups.

Polystyrene can be recycled, and has the number "6" as its recycling symbol. The increasing oil prices have increased the value of polystyrene for recycling. No known microorganism has yet been shown to biodegrade polystyrene, and it is often abundant as a form of pollution in the outdoor environment, particularly along shores and waterways especially in its low density cellular form.

### *History*

Polystyrene was discovered in 1839 by Eduard Simon, an apothecary in Berlin. From storax, the resin of the Turkish sweetgum tree Liquidambar orientalis, he distilled an oily substance, a monomer which he named styrol. Several days later, Simon found that the styrol had thickened, presumably from oxidation, into a jelly he dubbed styrol oxide ("Styroloxyd"). By 1845 English chemist John Blyth and German chemist August Wilhelm von Hofmann showed that the same transformation of styrol took place in the absence of oxygen. They called their substance metastyrol. Analysis later showed that it was chemically identical to Styroloxyd. In 1866 Marcelin Berthelot correctly identified the formation of metastyrol from styrol as a polymerization process. About 80 years went by before it was realized that heating of styrol starts a chain reaction which produces macromolecules, following the thesis of German organic chemist Hermann Staudinger (1881–1965). This eventually led to the substance receiving its present name, polystyrene.

The company I. G. Farben began manufacturing polystyrene in Ludwigshafen, Germany, about 1931, hoping it would be a suitable replacement for die-cast zinc in many applications. Success was achieved when they developed a reactor vessel that extruded polystyrene through a heated tube and cutter, producing polystyrene in pellet form. Before 1949, the chemical engineer Fritz Stastny (1908–1985) developed pre-expanded PS beads by incorporating aliphatic hydrocarbons, such as pentane. These beads are the raw material for moulding parts or extruding sheets. BASF and Stastny applied for a patent which was issued in 1949. The moulding process was demonstrated at the Kunststoff Messe 1952 in Düsseldorf. Products were named Styropor.

The crystal structure of isotactice polystyrene was reported by Giulio Natta.

In 1959, the Koppers Company in Pittsburgh, Pennsylvania, developed expanded polystyrene (EPS) foam.

### Structure

In chemical terms, polystyrene is a long chain hydrocarbon wherein alternating carbon centers are attached to phenyl groups (the name given to the aromatic ring benzene. Polystyrene's chemical formula is (C8H8)n; it contains the chemical elements carbon and hydrogen.

The materials properties are determined by short range van der Waals attractions between polymers chains. Since the molecules are long hydrocarbon chains that consist of thousands of atoms, the total attractive force between the molecules is large. When heated (or, equivalently, deformed at a rapid rate, due to a combination of viscoelastic and thermal insulation properties), the chains are able to take on a higher degree of conformation and slide past each other. This intermolecular weakness (versus the high intramolecular strength due to the hydrocarbon backbone) confers flexibility and elasticity. The ability of the system to be readily deformed above its glass transition temperature allows polystyrene (and thermoplastic polymers in general) to be readily softened and molded upon heating.

### Polymerization

Polystyrene results when styrene monomers interconnect. In the polymerization one carbon-carbon double bond (in the vinyl group) is replaced by a much stronger carbon-carbon single bond, hence it is very difficult to depolymerize polystyrene. Typically about a few thousand monomers comprise a chain of polystyrene, giving a molecular weight of 100,000–400,000.

A 3-D model would show that each of the chiral backbone carbons lies at the centre of a tetrahedron, with its 4 bonds pointing toward the vertices. Say, the -C-C- bonds are rotated so that the backbone chain lies entirely in the plane of the diagram. From this flat schematic, it is not evident which of the phenyl (benzene) groups are angled toward us from the plane of the diagram, and which ones are angled away. The isomer where all of them are on the same side is called isotactic polystyrene, which is not produced commercially.

### Atactic Polystyrene

The only commercially important form of polystyrene is atactic, which means that the phenyl groups are randomly distributed on both sides of the polymer chain. This random positioning prevents the chains from ever aligning with sufficient regularity to achieve any crystallinity. The plastic has a glass transition temperature Tg of ~90C. Polymerization is initiated with free radicals.

### Isotactic and Syndiotactic Polystyrene

Ziegler-Natta polymerization can produce an ordered syndiotactic polystyrene with the phenyl groups positioned on alternating sides of the hydrocarbon backbone. This form is highly crystalline with a Tm of 270 °C (518 °F). Such materials are not commercially produced because the polymerization is slow.

Extruded polystyrene is about as strong as an unalloyed aluminium, but much more flexible and much lighter (1.05 g/cm3 vs. 2.70 g/cm3 for aluminium).

### Degradation

Because it is an aromatic hydrocarbon, it burns with an orange-yellow flame, giving off soot, as is characteristic of materials containing aromatic rings. Complete oxidation of polystyrene produces carbon dioxide and water vapour. Because of its chemical inertness, polystyrene is used to fabricate containers for chemicals, solvents, and foods.

### Forms Produced

| ***Properties*** | |
|---|---|
| Density | 1.05 g/cm3 |
| Density of EPS | 16–640 kg/m3 |
| Dielectric constant | 2.4–2.7 |
| Electrical conductivity (s) | 10"16 S/m |
| Thermal conductivity (k) | 0.08 W/(m·K) |
| Young's modulus (E) | 3000–3600 MPa |
| Tensile strength (st) | 46–60 MPa |

| | |
|---|---|
| Elongation at break | 3–4% |
| Notch test | 2–5 kJ/m2 |
| Glass transition temperature | 95 °C |
| Melting point | 240 °C |
| Vicat B | 90 °C |
| Linear expansion coefficient (a) | 8×10"5/K |
| Specific heat (c) | 1.3 kJ/(kg·K) |
| Water absorption (ASTM) | 0.03–0.1 |
| Decomposition | X years, still decaying |

Polystyrene is commonly injection molded or extruded, while expanded polystyrene is either extruded or molded in a special process. Polystyrene copolymers are also produced; these contain one or more other monomers in addition to styrene. In recent years the expanded polystyrene composites with cellulose and starch have also been produced.

Extruded closed-cell polystyrene foam is sold under the trademark Styrofoam by Dow Chemical. This term is often used informally for other foamed polystyrene products.

Polystyrene is used in some polymer-bonded explosives:

| ***Polystyrene PBX examples Name*** | ***Explosive ingredients*** | ***Binder ingredients*** |
|---|---|---|
| PBX-9205 | RDX 92% | Polystyrene 6%; DOP 2% |
| PBX-9007 | RDX 90% | Polystyrene 9.1%; DOP 0.5%; resin 0.4% |

### *Sheet or Molded Polystyrene*

CD case made from general purpose polystyrene (GPPS) and high impact polystyrene (HIPS).

### *Disposable Polystyrene Razor*

Polystyrene (PS) is economical, and is used for producing plastic model assembly kits, plastic cutlery, CD “jewel” cases, smoke detector housings, license plate frames, and many other objects where a fairly rigid, economical plastic is desired. Production methods include thermoforming and injection molding.

Polystyrene Petri dishes and other laboratory containers such as test tubes and microplates play an important role in biomedical research and science. For these uses, articles are almost always made by injection molding, and often sterilized post-molding, either by irradiation or treatment with ethylene oxide. Post-mold surface modification, usually with oxygen-rich plasmas, is often done to introduce polar groups. Much of modern biomedical research relies on the use of such products; they therefore play a critical role in pharmaceutical research.

### Foams

Polystyrene foams are good thermal insulators and are therefore often used as building insulation materials, such as in structural insulated panel building systems. They are also used for non-weight-bearing architectural structures (such as ornamental pillars). PS foams exhibit also good damping properties, therefore it is used widely in packaging.

### Expanded Polystyrene

Expanded polystyrene (EPS) is a rigid and tough, closed-cell foam. It is usually white and made of pre-expanded polystyrene beads. Familiar uses include molded sheets for building insulation and packing material ("peanuts") for cushioning fragile items inside boxes. Sheets are commonly packaged as rigid panels (size 4 by 8 or 2 by 8 feet in the United States), which are also known as "bead-board". Thermal resistivity is usually about 28 m·K/W (or R-4 per inch). Some EPS boards have a flame spread of less than 25 and a smoke-developed index of less than 450, which means they can be used without a fire barrier (but require a 15 minute thermal barrier) according to US building codes. A growing use of EPS in construction is insulating concrete forms. The density range is about 16–640 kg/m3. The most common processing method is thermal cutting with hot wires.

### Extruded Polystyrene Foam

Extruded polystyrene foam (XPS) consists of closed cells, offers improved surface roughness and higher stiffness and reduced thermal conductivity. The density range is about 28 – 45 kg/m3.

Extruded polystyrene material is also used in crafts and model building, particularly architectural models. Because of the extrusion manufacturing process, XPS does not require facers to maintain its thermal or physical property performance. Thus, it makes a more uniform substitute for corrugated cardboard. Thermal resistivity is usually about 35 m·K/W (or R-5 per inch in American customary units).

Styrofoam is a trademarked name for XPS; however, it is often also used in the United States as a generic name for all polystyrene foams.

### *Copolymers*

Pure polystyrene is brittle, but hard enough that a fairly high-performance product can be made by giving it some of the properties of a stretchier material, such as polybutadiene rubber. The two such materials can never normally be mixed because of the amplified effect of intermolecular forces on polymer insolubility, but if polybutadiene is added during polymerization it can become chemically bonded to the polystyrene, forming a graft copolymer which helps to incorporate normal polybutadiene into the final mix, resulting in high-impact polystyrene or HIPS, often called "high-impact plastic" in advertisements. One commercial name for HIPS is Bextrene. Common applications of HIPS include toys and product casings. HIPS is usually injection molded in production. Autoclaving polystyrene can compress and harden the material.

Several other copolymers are also used with styrene. Acrylonitrile butadiene styrene or ABS plastic is similar to HIPS: a copolymer of acrylonitrile and styrene, toughened with polybutadiene. Most electronics cases are made of this form of polystyrene, as are many sewer pipes. SAN is a copolymer of styrene with acrylonitrile, and SMA one with maleic anhydride. Styrene can be copolymerized with other monomers; for example, divinylbenzene can be used for cross-linking the polystyrene chains to give the polymer used in Solid phase peptide synthesis.

### *Oriented Polystyrene*

Oriented polystyrene (OPS) is produced by stretching extruded PS film, improving visibility through the material by reducing haziness and increasing stiffness. This is often used in packaging where the manufacturer would like the consumer to see the enclosed product. Some benefits to OPS are that it is less expensive to produce than other clear plastics such as PP, PET and HIPS, and it is less hazy than HIPS or PP. The main disadvantage to OPS is that it's brittle. It will crack or tear easily.

### *Disposal and Environmental Issues*

Polystyrene is easily recycled. Due its light weight (especially if foamed) it is not economical to collect in its original form. However if the waste material goes through an initial compaction process the material changes density from typically 30 kg/m3 to 330 kg/m3 and becomes a recyclable commodity of high value for producers of recycled plastic pellets. It is generally not accepted in curbside collection recycling programs. In Germany polystyrene is collected, as a consequence of the packaging law (Verpackungsverordnung) that requires manufacturers to take responsibility for recycling or disposing of any packaging material they sell. In the US and many other countries the interest in recycling polystyrene has led to collection points being established. The producers of large quantities of polystyrene waste (50 tons per year or more) who have invested in the EPS compactors are able to sell the compacted blocks to plastic recyclers.

### Environmental Impact

Discarded polystyrene does not biodegrade for hundreds of years and is resistant to photolysis. Because of this stability, very little of the waste discarded in today's modern, highly engineered landfills biodegrades. Because degradation of materials creates potentially harmful liquid and gaseous by-products that could contaminate groundwater and air, today's landfills are designed to minimize contact with air and water required for degradation, thereby practically eliminating the degradation of waste.

Polystyrene foam is a major component of plastic debris in the ocean, where it becomes toxic to marine life. Foamed polystyrene blows in the wind and floats on water, and is abundant in the outdoor environment. Polystyrene foams are produced using blowing agents that form bubbles and expand the foam. In expanded polystyrene, these are usually hydrocarbons such as pentane, which may pose a flammability hazard in manufacturing or storage of newly manufactured material, but have relatively mild environmental impact. However, extruded polystyrene is usually made with hydrochlorofluorocarbons (HCFC) blowing agents which have effects on ozone depletion and on global warming. Their ozone depletion potential is greatly reduced relative to chlorofluorocarbon (CFC) which were formerly used, but their global warming potential can be on the order of 1000 or more, meaning it has 1000 times greater effect on global warming than does carbon dioxide.

In 2007, Eben Bayer, a student at Rensselaer Polytechnic Institute, invented an environmentally-friendly replacement for polystyrene packaging using the mineral perlite bonded with lignin-containing agricultural waste, degraded by fungi, which he calls Ecocradle.

### *Recycling*

The resin identification code symbol for polystyrene Currently, most polystyrene products are not recycled due to the lack of incentive to invest in the compactors and logistical systems required. Expanded polystyrene scrap can be easily added to products such as EPS insulation sheets and other EPS materials for construction applications. Commonly, manufacturers cannot obtain sufficient scrap because of the aforementioned collection issues. When it is not used to make more EPS, foam scrap can be turned into clothes hangers, park benches, flower pots, toys, rulers, stapler bodies, seedling containers, picture frames, and architectural molding from recycled PS.

Recycled EPS is also used in many metal casting operations. Rastra is made from EPS that is combined with cement to be used as an insulating amendment in the making of concrete foundations and walls. American manufacturers have produced insulating concrete forms made with approximately 80% recycled EPS since 1993. However, polystyrene recycling is not a closed loop, producing more polystyrene; polystyrene cups and other packaging materials are instead usually used as fillers in other plastics, or in other items that cannot themselves be recycled and are thrown away.

### *Incineration*

If polystyrene is properly incinerated at high temperatures, the chemicals generated are water, carbon dioxide, a complex mixture of volatile compounds, and carbon soot. According to the American Chemistry Council, when polystyrene is incinerated in modern facilities, the final volume is 1% of the starting volume; most of the polystyrene is converted into carbon dioxide, water vapour, and heat. Because of the amount of heat released, it is sometimes used as a power source for steam or electricity generation.

When polystyrene was burned at temperatures of 800–900 °C (the typical range of a modern incinerator), the products of combustion consisted of “a complex mixture of polycyclic aromatic hydrocarbons (PAHs) from alkyl benzenes to benzoperylene. Over 90 different compounds were identified in combustion effluents from polystyrene.”

When burned without enough oxygen or at lower temperatures (as in a campfire or a household fireplace), polystyrene can produce polycyclic aromatic hydrocarbons, carbon black, and carbon monoxide, as well as styrene monomers.

### Burial

This section does not cite any references or sources. Please help improve this section by adding citations to reliable sources. Unsourced material may be challenged and removed.

The neutrality of this article is disputed.

Foam cups and other polystyrene products can be safely buried in landfills, since it is as stable as concrete or brick. No plastic film is required to protect the air and underground water.

### Reducing

Some effort is being made to find alternatives to polystyrene foam, especially in restaurant settings. Restricting the use of foamed polystyrene takeout food packaging is a priority of many solid waste environmental organizations. However, the Plastics Foodservice Packaging Group counters that in US, less than 1% by weight of solid waste disposed is polystyrene. A campaign to achieve the first ban of polystyrene foam from the food & beverage industry in Canada was launched in Toronto as of January 2007, by local non-profit organization NaturoPack. Portland, Ore., and San Francisco are among about one hundred cities in the United States that currently have some sort of ban on polystyrene foam in restaurants. For instance, in 2007 restaurants in Oakland, California were required to switch to disposable food containers that will biodegrade if added to food compost.

Although polystyrene can be recycled at recycling facilities, most polystyrene is not recycled. The EPA (United States Environmental Protection Agency) estimates that 25 billion polystyrene cups are tossed every year. Since polystyrene degrades very slowly- more than 500 years for a single cup– the EPA considers this a serious environmental problem. Several green leaders, from the Dutch Ministry of the Environment to Starbucks' Green Team, advise that individuals reduce their environmental impact by using reusable coffee cups.

### Finishing

This section does not cite any references or sources. Please help improve this section by adding citations to reliable sources. Unsourced material may be challenged and removed. In the United States, environmental protection regulations prohibit the use of solvents on polystyrene (which would dissolve the polystyrene and de-foam most foams anyway).

Some acceptable finishing materials are

* Water-based paint (artists have created paintings on polystyrene with gouache)
* Mortar or acrylic/cement render, often used in the building industry as a weather-hard overcoat that hides the foam completely after finishing the objects.
* Cotton wool or other fabrics used in conjunction with a stapling implement.

### Safety—Health

Based on scientific tests over five decades, government safety agencies have determined that polystyrene is safe for use in foodservice products. For example, polystyrene meets the stringent standards of the U.S. Food and Drug Administration and the European Commission/ European Food Safety Authority for use in packaging to store and serve food. The Hong Kong Food and Environmental Hygiene Department recently reviewed the safety of serving various foods in polystyrene foodservice products and reached the same conclusion as the U.S. FDA.

From 1999 to 2002, a comprehensive review of the potential health risks associated with exposure to styrene was conducted by a 12 member international expert panel selected by the Harvard Centre for Risk Assessment. The scientists had expertise in toxicology, epidemiology, medicine, risk analysis, pharmacokinetics, and exposure assessment.

The Harvard study reported that styrene is naturally present in foods such as strawberries, beef, and spices, and is naturally produced in the processing of foods such as wine and cheese. The study also reviewed all the published data on the quantity of styrene contributing to the diet due to migration of food packaging and disposable food contact articles, and concluded there is no cause for concern for the general public from exposure to styrene from foods or styrenic materials used in food-contact applications, such as polystyrene packaging and food service containers.

Styrene oligomers in polystyrene containers used for food packaging have been found to migrate into the food. Another Japanese study conducted on wild-type and AhR-null mice found that the styrene trimer, which the authors detected in cooked polystyrene container-packed instant foods, may increase thyroid hormone levels.

### *Fire Hazards*

Like other organic compounds, polystyrene is flammable. Polystyrene is classified according to DIN4102 as a "B3" product, meaning highly flammable or "easily ignited." Consequently, although it is an efficient insulator at low temperatures, its use is prohibited in any exposed installations in building construction if the material is not flame retardant. It must be concealed behind drywall, sheet metal or concrete. Foamed polystyrene plastic materials have been accidentally ignited and caused huge fires and losses, for example at the Düsseldorf International Airport, the Channel tunnel (where polystyrene was inside a railcar that caught fire), and the Browns Ferry Nuclear Power Plant (where fire breached a fire retardant and reached the foamed plastic underneath, inside a firestop that had not been tested and certified in accordance with the final installation).

## Cumene

Cumene is the common name for isopropylbenzene, an organic compound that is an aromatic hydrocarbon. It is a constituent of crude oil and refined fuels. It is a flammable colorless liquid that has a boiling point of 152 °C. Nearly all the cumene that is produced as a pure compound on an industrial scale is converted to cumene hydroperoxide, which is an intermediate in the synthesis of other industrially important chemicals, primarily phenol and acetone.

### *Production*

Commercial production of cumene is carried out through the catalytic alkylation of benzene, with the addition of propylene. Solid phosphoric acid (SPA) supported on alumina can also be used as a catalyst, and this was the case prior to the mid-1990s when zeolite-based catalysts made the other technique commercially redundant. This process is other wise known as Friedel-Crafts alkylation reaction.

Isopropylbenzene is stable, but may form peroxides in storage if in contact with the air. It is important to test for the presence of peroxides before heating or distilling. The chemical is also flammable and incompatible with strong oxidizing agents. Environmental laboratories commonly test isopropylbenzene using a GCMS instrument.

## Cumene Process

The Cumene process (Cumene-phenol process, Hock process) is an industrial process for developing phenol and acetone from benzene and propylene. The term stems from cumene (isopropyl benzene), the intermediate material during the process. It was invented by Heinrich Hock in 1944 and independently by R. Ûdris and P. Sergeyev in 1942 (USSR).

This process converts two relatively cheap starting materials, benzene and propylene, into two more valuable ones, phenol and acetone. Other reactants required are oxygen from air and small amounts of a radical initiator. Most of the worldwide production of phenol and acetone is now based on this method. In 2003, nearly 7 billion kg of phenol was produced by the Hock Process.

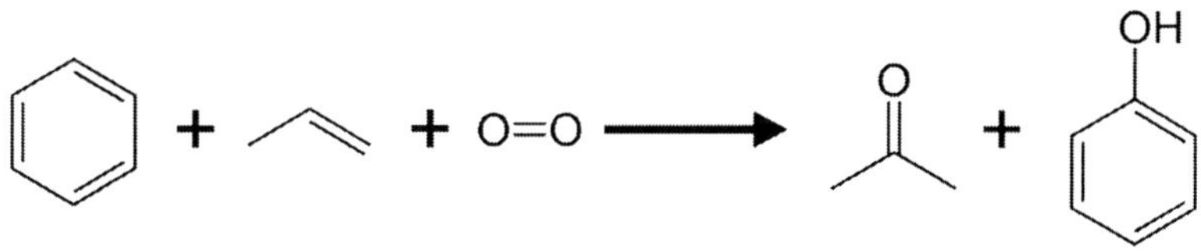


### *Technical Description*

Benzene and propylene are compressed together to a pressure of 30 standard atmospheres at 250 °C (482 °F) in presence of a catalytic Lewis acid. Phosphoric acid is often favoured over aluminium halides. Cumene is formed in the gas-phase Friedel-Crafts alkylation of benzene by propylene:

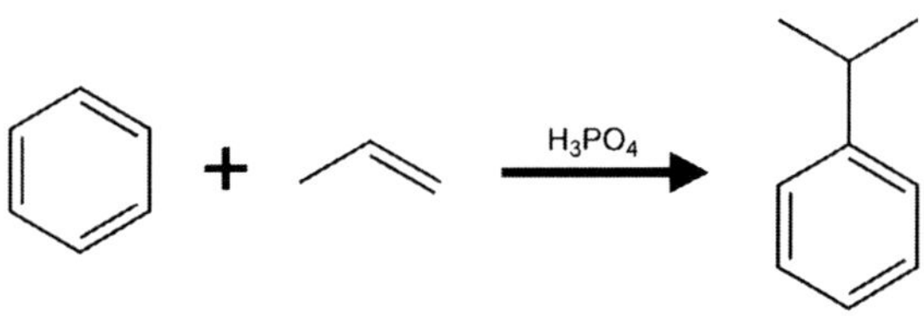


Cumene is oxidized in air which removes the tertiary benzylic hydrogen from cumene and hence forms a cumene radical:

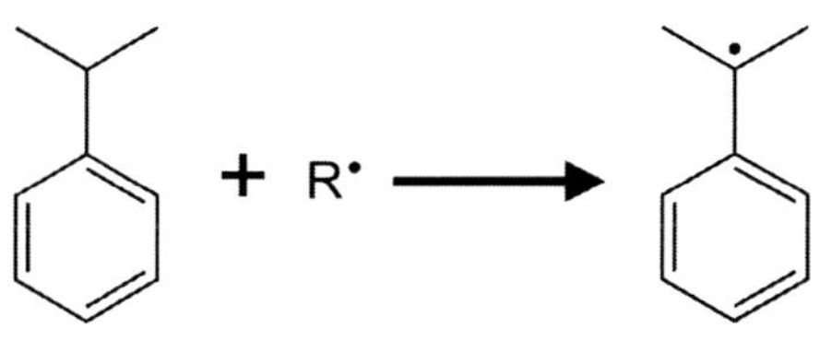

This cumene radical then bonds with an oxygen molecule to give cumene hydroperoxide radical, which in turn forms cumene hydroperoxide ($C_6H_5C(CH_3)_2$-O-O-H) by abstracting benzylic hydrogen from another cumene molecule. This latter cumene converts into cumene radical and feeds back into subsequent chain formations of cumene hydroperoxides. A pressure of 5 atm is used to ensure that the unstable peroxide is kept in liquid state.

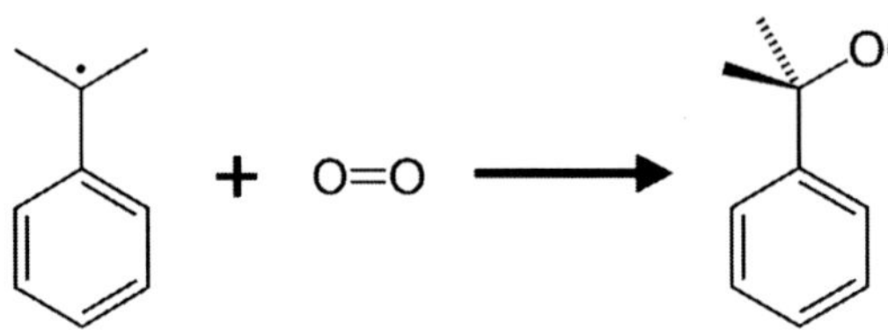

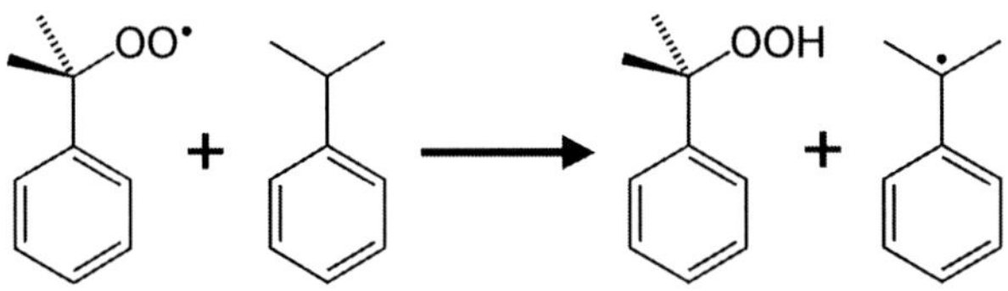

Cumene hydroperoxide is then hydrolysed in an acidic medium (the Hock rearrangement) to give phenol and acetone. In the first step, the terminal hydroperoxy oxygen atom is protonated. This is followed by a step in which the phenyl group migrates from the benzyl carbon to the adjacent oxygen and a water molecule is lost, producing a resonance stabilized tertiary carbocation. The concerted mechanism of this step is similar to the mechanisms of the Baeyer-Villiger oxidation and also the oxidation step of hydroboration-oxidation. In 2009, an acidified bentonite clay was proven to be a more economical catalyst than sulfuric acid as the acid medium.

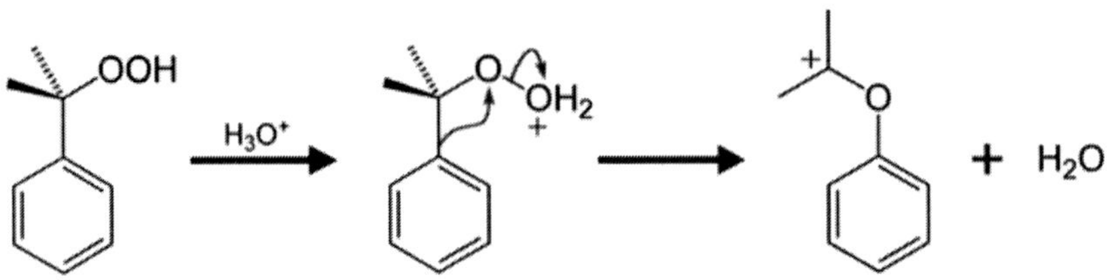

As shown below, the resulting carbocation is then attacked by water, a proton is then transferred from the hydroxy oxygen to the ether oxygen, and finally the ion falls apart into phenol and acetone.

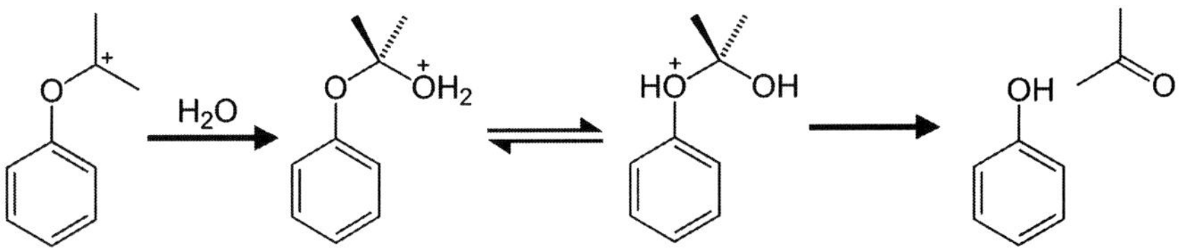


The products are extracted by distillation.

## Phenol

Phenol, also known as carbolic acid, is an organic compound with the chemical formula $C_6H_5OH$. It is a white crystalline solid. The molecule consists of a phenyl ($-C_6H_5$), bonded to a hydroxyl (-OH) group. It is produced on a large scale (about 7 billion kg/year) as a precursor to many materials and useful compounds. It is only mildly acidic but requires careful handling due to its propensity to cause burns.

Phenol was first extracted from coal tar, and its major uses involve its conversion to plastics or related materials. Phenols are key for building polycarbonates, epoxies, Bakelite, nylon, detergents and a large collection of drugs, herbicides and pharmaceuticals.

### *Properties*

Phenol is appreciably soluble in water, with about 8.3 g dissolving in 100 mL (0.88 M). The sodium salt of phenol, sodium phenoxide, is far more water soluble.

### *Acidity*

It is slightly acidic: the phenol molecules have weak tendencies to lose the $H^+$ ion from the hydroxyl group, resulting in the highly water-soluble phenolate anion $C_6H_5O^-$ (also called phenoxide). Compared to aliphatic alcohols, phenol is about 1 million times more acidic, although it is still considered a weak acid. It reacts completely with aqueous NaOH to lose $H^+$, whereas most alcohols react only partially. Phenols are less acidic than carboxylic acids, and even carbonic acid.

One explanation for the increased acidity over alcohols is resonance stabilization of the phenoxide anion by the aromatic ring. In this way, the negative charge on oxygen is shared by the ortho and para carbon atoms. In another explanation, increased acidity is the result of orbital overlap between the oxygen's lone pairs and the aromatic system.

In a third, the dominant effect is the induction from the $sp^2$ hybridised carbons; the comparatively more powerful inductive withdrawal of electron density that is provided by the $sp^2$ system compared to an $sp^3$ system allows for great stabilization of the oxyanion.

In making this conclusion, one can examine the $pK_a$ of the enol of acetone, which is 19.0, in comparison to phenol with a $pK_a$ of 10.0. However, this similarity of acidities of phenol and acetone enol is not observed in the gas phase, and is because the difference of solvation energies of the deprotonated acetone enol and phenoxide almost exactly offsets the experimentally observed gas phase acidity difference. It has recently been shown that only about 1/3 of the increased acidity of phenol is due to inductive effects, with resonance accounting for the rest.

### *Phenoxide Anion*

Phenol can be deprotonated with moderate base such as triethylamine, forming the nucleophilic *phenoxide anion* or *phenolate anion*, which is highly water-soluble.

The phenoxide anion has a similar nucleophilicity to free amines, with the further advantage that its conjugate acid (neutral phenol) does not become entirely deactivated as a nucleophile even in moderately acidic conditions. Phenols are sometimes used in peptide synthesis to "activate" carboxylic acids or esters to form activated esters. Phenolate esters are far more stable than acid anhydrides or acyl halides but are sufficiently reactive under mild conditions to facilitate the formation of amide bonds.

Phenoxides are enolates stabilised by aromaticity. Under normal circumstances, phenoxide is more reactive at the oxygen position, but the oxygen position is a "hard" nucleophile whereas the alpha-carbon positions tend to be "soft".

### *Tautomerism*

Phenol exhibits keto-enol tautomerism with its unstable keto tautomer cyclohexadienone, but only a tiny fraction of phenol exists as the keto form. The equilibrium constant for enolisation is approximately 10, meaning that only one in every ten trillion molecules is in the keto form at any moment. The small amount of stabilisation gained by exchanging a C=C bond for a C=O bond is more than offset by the large destabilisation resulting from the loss of aromaticity. Phenol therefore exists entirely in the enol form.

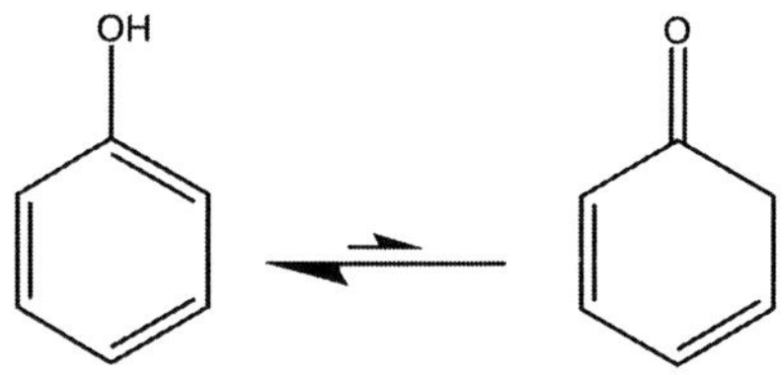


*Phenol-cyclohexadienone Tautomerism*

### Reactions

Phenol is highly reactive toward electrophilic aromatic substitution as the oxygen atom's pi electrons donate electron density into the ring. By this general approach, many groups can be appended to the ring, via halogenation, acylation, sulfonation, and other processes. However, phenol's ring is so strongly activated - second only to aniline - that bromination or chlorination of phenol leads to substitution on all carbons ortho and para to the hydroxy group, not only on one carbon.

### Production

Because of phenol's commercial importance, many methods have been developed for its production. The main route, accounting for 95% of production (2003), involves the partial oxidation of cumene (isopropylbenzene) via the Hock rearrangement:

$$C_6H_5CH(CH_3)_2 + O_2 \rightarrow C_6H_5OH + (CH_3)_2CO$$

An early commercial route begins with the reaction of strong base with benzenesulfonate:

$$C_6H_5SO_3H + 2\ NaOH \rightarrow C_6H_5OH + Na_2SO_3 + H_2O$$

Other methods under consideration involve:

- hydrolysis of chlorobenzene, using base or steam (Raschig-Hooker process):

$$C_6H_5Cl + H_2O \rightarrow C_6H_5OH + HCl$$

- direct oxidation of benzene with nitrous oxide, a potentially "green" process:

$$C_6H_6 + N_2O \rightarrow C_6H_5OH + N_2$$

- oxidation of toluene, as developed by Dow Chemical:

$$C_6H_5CH_3 + 2\ O_2 \rightarrow C_6H_5OH + CO_2 + H_2O$$

In the Lummus Process, the oxidation of toluene to benzoic acid is conducted separately.

Phenol is also recoverable byproduct of coal pyrolysis.

### Uses

The major uses of phenol, consuming two thirds of its production, involve its conversion to plastics or related materials. Condensation with acetone gives bisphenol-A, a key precursor to polycarbonates and epoxide resins. Condensation with formaldehyde gives phenolic resins, a famous example of which is Bakelite. Hydrogenation of phenol gives cyclohexanone, a precursor to nylon. Nonionic detergents are produced by alkylation of phenol to give the alkylphenols, e.g., nonylphenol, which are then subjected to ethoxylation.

Phenol is also a versatile precursor to a large collection of drugs, most notably aspirin but also many herbicides and pharmaceuticals. Phenol is also used as an oral anesthetic/analgesic, commonly used to temporarily treat pharyngitis.

### Niche Uses

Phenol is so inexpensive that it attracts many small-scale uses. It once was widely used as an antiseptic, especially as Carbolic soap, from the early 1900s through the 1970s. It is a component of industrial paint strippers used in the aviation industry for the removal of epoxy, polyurethane and other chemically resistant coatings.

Phenol derivatives are also used in the preparation of cosmetics including sunscreens, hair dyes, and skin lightening preparations.

### History

Phenol was discovered in 1834, when it was extracted from coal tar, which remained the primary source until the development of the petrochemical industry.

The antiseptic properties of phenol were used by Sir Joseph Lister (1827–1912) in his pioneering technique of antiseptic surgery, although the skin irritation caused by continual exposure to phenol eventually led to the substitution of aseptic (germ-free) techniques in surgery. Lister decided that the wounds themselves had to be thoroughly cleaned. He then covered the wounds with a piece of rag or lint covered in phenol, or carbolic acid as he called it. It is also the active ingredient in some oral analgesics such as Chloraseptic spray as well as Carmex. Phenol was also the main ingredient of the Carbolic Smoke Ball, an ineffective device marketed in London in the 19th century as protecting the user against influenza and other ailments, and the subject of a famous law case.

### Second World War

Injections of phenol have occasionally been used as a means of execution. In particular, phenol and cyanide injections were used as a means of individual execution by the Nazis during the Second World War. Originally used by the Nazis in 1939 as part of Action T4, phenol, inexpensive, easy to make and quickly lethal, became the injectable toxin of choice as part of Nazi Germany's "euthanasia" program. Although Zyklon-B pellets, invented by Gerhard Lenz, were used in the gas chambers to exterminate large groups of people, the Nazis learned that extermination of smaller groups was more economical via injection of each victim, one at a time, with phenol. Phenol injections were given to thousands of people in concentration camps, especially at Auschwitz-Birkenau. Approximately one gram is enough to cause death. Injections were administered by medical doctors, their assistants, or sometimes prisoner doctors; such injections were originally given intravenously, more commonly in the arm, but injection directly into the heart, so as to induce nearly instant death, was later adopted. One of the best known inmates to be executed with a phenol injection in Auschwitz was St. Maximilian Kolbe, a Catholic priest who volunteered to undergo three weeks of starvation and dehydration in the place of another inmate.

### Occurrence

Phenol is a measurable component in the aroma and taste of the distinctive Islay scotch whisky, generally ~30, but up to 100 ppm.

### Toxicity

Phenol and its vapours are corrosive to the eyes, the skin, and the respiratory tract. Repeated or prolonged skin contact with phenol may cause dermatitis, or even second and third-degree burns due to phenol's caustic and defatting properties. Inhalation of phenol vapour may cause lung edema. The substance may cause harmful effects on the central nervous system and heart, resulting in dysrhythmia, seizures, and coma. The kidneys may be affected as well. Exposure may result in death and the effects may be delayed. Long-term or repeated exposure of the substance may have harmful effects on the liver and kidneys." There is no evidence to believe that phenol causes cancer in humans. Besides its hydrophobic effects, another mechanism for the toxicity of phenol may be the formation of phenoxyl radicals.

Chemical burns from skin exposures can be decontaminated by washing with polyethylene glycol, isopropyl alcohol, or perhaps even copious amounts of water. Removal of contaminated clothing is required, as well as immediate hospital treatment for large splashes. This is particularly important if the phenol is mixed with chloroform (a commonly-used mixture in molecular biology for DNA & RNA purification from proteins).

### *Phenols*

The word *phenol* is also used to refer to any compound that contains a six-membered aromatic ring, bonded directly to a hydroxyl group (-OH). Thus, phenols are a class of organic compounds of which the phenol discussed in this article is the simplest member.

## Bisphenol A

Bisphenol A (BPA) is an organic compound with two phenol functional groups. It is used to make polycarbonate plastic and epoxy resins, along with other applications.

As it has been known to be estrogenic since the mid 1930s, concerns about the use of bisphenol A in consumer products were regularly reported in the news media in 2008, after several governments issued reports questioning its safety, prompting some retailers to remove products containing it from their shelves. A 2010 report from the United States Food and Drug Administration (FDA) raised further concerns regarding exposure of fetuses, infants and young children. In September 2010, Canada became the first country to declare BPA as a toxic substance. In the European Union and Canada, BPA use is banned in baby bottles.

### *Production*

World production capacity of this compound was 1 million tons in the 1980s, and more than 2.2 million tons in 2009. In 2003, U.S. consumption was 856,000 tons, 72% of which was used to make polycarbonate plastic and 21% going into epoxy resins. In the US, less than 5% of the BPA produced is used in food contact applications.

Bisphenol A was first synthesized by the Russian chemist A.P. Dianin in 1891. This compound is synthesized by the condensation of acetone (hence the suffix A in the name) with two equivalents of phenol. The reaction is catalysed by a strong acid, such as hydrochloric acid (HCl) or a sulfonated polystyrene resin. Industrially, a large excess of phenol is used to ensure full condensation; the product mixture of the cumene process (acetone and phenol) may also be used as starting material:

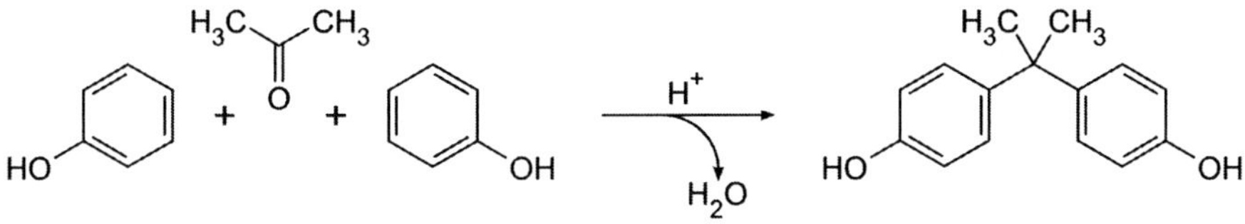

A large number of ketones undergo analogous condensation reactions. Commercial production of BPA requires distillation – either extraction of BPA from many resinous byproducts under high vacuum, or solvent-based extraction using additional phenol followed by distillation.

### Use

Bisphenol A is used primarily to make plastics, and products containing bisphenol A-based plastics have been in commerce use since 1957. At least 8 billion pounds of BPA are used by manufacturers yearly. It is a key monomer in production of epoxy resins and in the most common form of polycarbonate plastic. The overall reaction to give polycarbonate can be written:

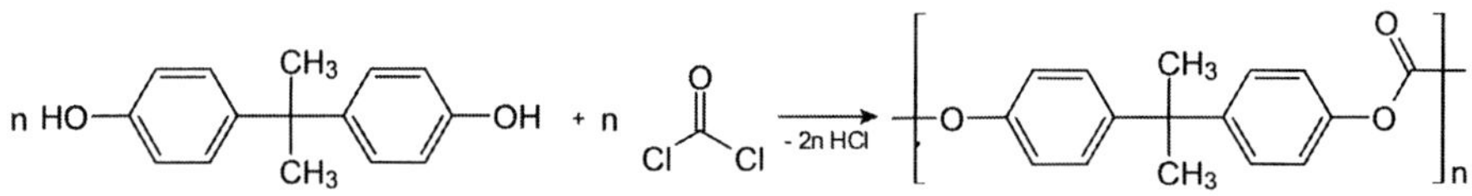

Polycarbonate plastic, which is clear and nearly shatter-proof, is used to make a variety of common products including baby and water bottles, sports equipment, medical and dental devices, dental fillings and sealants, eyeglass lenses, CDs and DVDs, and household electronics. BPA is also used in the synthesis of polysulfones and polyether ketones, as an antioxidant in some plasticizers, and as a polymerization inhibitor in PVC. Epoxy resins containing bisphenol A are used as coatings on the inside of almost all food and beverage cans, however, due to BPA health concerns, in Japan epoxy coating was mostly replaced by PET film. Bisphenol A is also a precursor to the flame retardant tetrabromobisphenol A, and was formerly used as a fungicide. Bisphenol A is a preferred colour developer in carbonless copy paper and thermal paper, with the most common public exposure coming from some thermal point of sale receipt paper. BPA-based products are also used in foundry castings and for lining water pipes.

### *Identification in Plastics*

"In general, plastics that are marked with recycle codes 1, 2, 4, 5, and 6 are very unlikely to contain BPA. Some, but not all, plastics that are marked with recycle codes 3 or 7 may be made with BPA."

There are seven classes of plastics used in packaging applications. Type 7 is the catch-all "other" class, and some type 7 plastics, such as polycarbonate (sometimes identified with the letters "PC" near the recycling symbol) and epoxy resins, are made from bisphenol A monomer.

Type 3 (PVC) can also contain bisphenol A as an antioxidant in plasticizers. This is particularly true for "flexible PVC", but not true for PVC pipes.

### *Health Effects*

Bisphenol A is an endocrine disruptor, which can mimic the body's own hormones and may lead to negative health effects. Early development appears to be the period of greatest sensitivity to its effects, and some studies have linked prenatal exposure to later neurological difficulties. Regulatory bodies have determined safety levels for humans, but those safety levels are currently being questioned or under review as a result of new scientific studies. A 2011 study that investigated the number of chemicals pregnant women are exposed to in the U.S. found BPA in 96% of women.

In 2009, The Endocrine Society released a statement expressing concern over current human exposure to BPA.

In 2011, the Food Standards Agency's chief scientist said "the evidence [is] that BPA is rapidly absorbed, detoxified, and eliminated from humans – therefore is not a health concern."

### *Expert Panel Conclusions*

In 2007, a consensus statement by 38 experts on bisphenol A concluded that average levels in people are above those that cause harm to many animals in laboratory experiments. However, they noted that while BPA is not persistent in the environment or in humans, biomonitoring surveys indicate that exposure is continuous, which is problematic because acute animal exposure studies are used to estimate daily human exposure to BPA, and no studies that had examined BPA pharmacokinetics in animal models had followed continuous low level exposures. They added that measurement of BPA levels in serum and other body fluids suggests that either BPA intake is much higher than accounted for, or that BPA can bioaccumulate in some conditions such as pregnancy, or both. A 2011 study, the first to examine BPA in a continuous low level exposure throughout the day, did find an increased absorption and accumulation of BPA in the blood of mice.

In 2007 it was reported that among government-funded BPA experiments on lab animals and tissues, 153 found adverse effects and 14 did not, whereas all 13 studies funded by chemical corporations reported no harm. The studies indicating harm reported a variety of deleterious effects in rodent offspring exposed in the womb: abnormal weight gain, insulin resistance, prostate cancer, and excessive mammary gland development.

A panel convened by the U.S. National Institutes of Health in 2007 determined that there was "some concern" about BPA's effects on fetal and infant brain development and behaviour. The concern over the effect of BPA on infants was also heightened by the fact that infants and children are estimated to have the highest daily intake of BPA. A 2008 report by the U.S. National Toxicology Program (NTP) later agreed with the panel, expressing "some concern for effects on the brain, behaviour, and prostate gland in fetuses, infants, and children at current human exposures to bisphenol A," and "*minimal* concern for effects on the mammary gland and an earlier age for puberty for females in fetuses, infants, and children at current human exposures to bisphenol A." The NTP had "*negligible* concern that exposure of pregnant women to bisphenol A will result in fetal or neonatal mortality, birth defects, or reduced birth weight and growth in their offspring."

### *Obesity*

A 2008 review has concluded that obesity may be increased as a function of BPA exposure, which "...merits concern among scientists and public health officials." A 2009 review of available studies has concluded that "perinatal BPA exposure acts to exert persistent effects on body weight and adiposity". Another 2009 review has concluded that "Eliminating exposures to (BPA) and improving nutrition during development offer the potential for reducing obesity and associated diseases". Other reviews have come with similar conclusions. A later study on rats has suggested that perinatal exposure to drinking water containing 1 mg/L of BPA increased adipogenesis in females at weaning. Other study suggested that larger size-for-age was due to a faster growth rate rather than obesity.

### Neurological Issues

A panel convened by the U.S. National Institutes of Health determined that there was "some concern" about BPA's effects on fetal and infant brain development and behaviour. A 2008 report by the U.S. National Toxicology Program (NTP) later agreed with the panel, expressing "some concern for effects on the brain". In January 2010 the FDA expressed the same level of concern.

A 2007 review has concluded that BPA, like other xenoestrogens, should be considered as a player within the nervous system that can regulate or alter its functions through multiple pathways. A 2007 review has concluded that low doses of BPA during development have persistent effects on brain structure, function and behaviour in rats and mice.

A 2008 review concluded that low-dose BPA maternal exposure causes long-term consequences at the level of neurobehavioural development in mice. A 2008 review has concluded that neonatal exposure to Bisphenol-A (BPA) can affect sexually dimorphic brain morphology and neuronal adult phenotypes in mice. A 2008 review has concluded that BPA altered long-term potentiation in the hippocampus and even nanomolar dosage could induce significant effects on memory processes. A 2009 review raised concerns about BPA effect on anteroventral periventricular nucleus.

A 2008 study by the Yale School of Medicine demonstrated that adverse neurological effects occur in non-human primates regularly exposed to bisphenol A at levels equal to the United States Environmental Protection Agency's (EPA) maximum safe dose of 50 µg/kg/day. This research found a connection between BPA and interference with brain cell connections vital to memory, learning and mood.

A 2010 study with rats prenatally exposed to 40 microg/kg bw BPA has concluded that corticosterone and its actions in the brain are sensitive to the programming effects of BPA.

### Disruption of the Dopaminergic System

A 2005 review concluded that prenatal and neonatal exposure to BPA in mice can potentiate the central dopaminergic systems, resulting in the supersensitivity to the drugs-of-abuse-induced reward effects and hyperlocomotion.

A 2008 review has concluded that BPA mimics estrogenic activity and impacts various dopaminergic processes to enhance mesolimbic dopamine activity resulting in hyperactivity, attention deficits, and a heightened sensitivity to drugs of abuse.

A 2009 study on rats has concluded that prenatal and neonatal exposure to low-dose BPA causes deficits in development at dorsolateral striatum via altering the function of dopaminergic receptors. Another 2009 study has found associated changes in the dopaminergic system.

### Thyroid Function

A 2007 review has concluded that bisphenol-A has been shown to bind to thyroid hormone receptor and perhaps have selective effects on its functions.

A 2009 review about environmental chemicals and thyroid function, raised concerns about BPA effects on triiodothyronine and concluded that "available evidence suggests that governing agencies need to regulate the use of thyroid-disrupting chemicals, particularly as such uses relate exposures of pregnant women, neonates and small children to the agents".

A 2009 review summarized BPA adverse effects on thyroid hormone action.

### Cancer Research

According to the WHO's INFOSAN, "animal studies have not provided convincing evidence of risk of cancer from BPA exposure."

Neither the U.S. Environmental Protection Agency nor the International Agency for Research on Cancer has evaluated bisphenol A for possible carcinogenic activity.

A 2010 review at Tufts University Medical School concluded that Bisphenol A may increase cancer risk.

### Breast Cancer

A 2008 review has concluded that "perinatal exposure to (...) low doses of (..) BPA, alters breast development and increases breast cancer risk". Another 2008 review concluded that "animal experiments and epidemiological data strengthen the hypothesis that fetal exposure to xenoestrogens may be an underlying cause of the increased incidence of breast cancer observed over the last 50 years".

A 2009 in vitro study has concluded that BPA is able to induce neoplastic transformation in human breast epithelial cells. Another 2009 study concluded that maternal oral exposure to low concentrations of BPA during lactation increases mammary carcinogenesis in a rodent model.

A 2010 study with the mammary glands of the offspring of pregnant rats treated orally with 0, 25 or 250 μg BPA/kg body weight has found that key proteins involved in signaling pathways such as cellular proliferation were regulated at the protein level by BPA.

A 2010 study has found that BPA may reduce sensitivity to chemotherapy treatment of specific tumours.

### *Neuroblastoma*

In vitro studies have suggested that BPA can promote the growth of neuroblastoma cells. A 2010 in vitro study has concluded that BPA potently promote invasion and metastasis of neuroblastoma cells through overexpression of MMP-2 and MMP-9 as well as downregulation of TIMP2.

### *Prostate Development and Cancer*

A 1997 study in mice has found that neonatal BPA exposure of 2 μg/kg increased adult prostate weight. A 2005 study in mice has found that neonatal BPA exposure at 10 μg/kg disrupted the development of the fetal mouse prostate.

A 2006 study in rats has shown that neonatal bisphenol A exposure at 10 μg/kg levels increases prostate gland susceptibility to adult-onset precancerous lesions and hormonal carcinogenesis. A 2007 in vitro study has found that BPA within the range of concentrations currently measured in human serum is associated with permanent increases in prostate size. A 2009 study has found that newborn rats exposed to a low-dose of BPA (10 μg/kg) increased prostate cancer susceptibility when adults.

### *DNA Methylation*

At least one study has suggested that bisphenol A suppresses DNA methylation which is linked to epigenetic changes.

### *Reproductive System and Sexual Behaviour Research*

A 2007 study using pregnant mice showed that BPA changes the expression of key developmental genes that form the uterus, which may impact female reproductive tract development and future fertility of female fetuses.

A series of studies made in 2009 found:

- Mouse ovary anomalies from exposure as low as 1 µg/kg, concluded that BPA exposure causes long-term adverse reproductive and carcinogenic effects if exposure occurs during prenatal critical periods of differentiation.
- Neonatal exposure of as low as 50 µg/kg disrupts ovarian development in mice.
- Neonatal BPA exposition of as low as 50 µg/kg permanently alters the hypothalamic estrogen-dependent mechanisms that govern sexual behaviour in the adult female rat.
- Prenatal exposure to BPA at levels of (10 µg/kg/day) affects behavioural sexual differentiation in male monkeys.
- In placental JEG3 cells *in vitro* BPA may reduce estrogen synthesis.
- BPA exposure disrupted the blood-testis barrier when administered to immature, but not to adult, rats.
- Exposure to BPA in the workplace was associated with self-reported adult male sexual dysfunction.

A 2009 rodent study, funded by EPA and conducted by some of its scientists, concluded that, compared with ethinyl estradiol, low-dose exposures of bisphenol A (BPA) showed no effects on several reproductive functions and behavioural activities measured in female rats. That study was criticized as flawed for using polycarbonate cages in the experiment (since polycarbonate contains BPA) and the claimed resistance of the rats to estradiol, but that claim was contested by the authors and others. Another 2009 rodent study found that BPA exposure during pregnancy has a lasting effect on one of the genes that is responsible for uterine development and subsequent fertility in both mice and humans (HOXA10). The authors concluded, "We don't know what a safe level of BPA is, so pregnant women should avoid BPA exposure."

In a 2010 study mice were given BPA at doses thought to be equivalent to levels currently being experienced by humans. The research showed that BPA exposure affects the earliest stages of egg production in the ovaries of the developing mouse fetuses, thus suggesting that the next generation may suffer genetic defects in such biological processes as mitosis and DNA replication.

In addition, the research team noted that their study "revealed a striking down-regulation of mitotic/cell cycle genes, raising the possibility that BPA exposure immediately before meiotic entry might act to shorten the reproductive lifespan of the female" by reducing the total pool of fetal oocytes. Another 2010 study with mice concluded that BPA exposure in utero leads to permanent DNA alterations in sensitivity to estrogen. Also in 2010, a rodent study found that by exposing fetal mice to BPA during pregnancy and examining gene expression and DNA in the uteruses of female fetuses, BPA exposure permanently affected the uterus by decreasing regulation of gene expression.

The changes caused the mice to over-respond to estrogen throughout adulthood, long after the BPA exposure, thus suggesting that early exposure to BPA genetically "programmed" the uterus to be hyper-responsive to estrogen. Extreme estrogen sensitivity can lead to fertility problems, advanced puberty, altered mammary development and reproductive function, as well as a variety of hormone-related cancers. One of the authors concluded that BPA may be similar to diethylstilbestrol caused birth defects and cancers in young women whose mothers were given the drug during pregnancy.

A 2011 study using the rhesus monkey, a species that is very similar to humans in regard to pregnancy and fetal development, found that prenatal exposure to BPA causes changes in female primates' uterus development. A 2011 rodent study found that male rats exposed to BPA had lower sperm counts and testosterone levels than those of unexposed males. A 2011 mice study found that male mice exposed to BPA became demasculinized and behaved more like females in their spatial navigational abilities. They were also less desirable to female mice.

### General Research

At an Endocrine Society meeting in 2009, new research reported data from animals experimentally treated with BPA. Studies presented at the group's annual meeting show BPA can affect the hearts of women, can permanently damage the DNA of mice, and appear to be entering the human body from a variety of unknown sources.

A 2009 *in vitro* study on cytotrophoblasts cells has found cytoxic effects in exposure of BPA doses from 0.0002 to 0.2 micrograms per millilitre and concluded this finding "suggests that exposure of placental cells to low doses of BPA may cause detrimental effects, leading *in vivo* to adverse pregnancy outcomes such as preeclampsia, intrauterine growth restriction, prematurity and pregnancy loss"

A 2009 study in rats concluded that BPA, at the reference safe limit for human exposure, was found to impact intestinal permeability and may represent a risk factor in female offspring for developing severe colonic inflammation in adulthood.

A 2010 study on mice has concluded that perinatal exposure to 10 micrograms/mL of BPA in drinking water enhances allergic sensitization and bronchial inflammation and responsiveness in an animal model of asthma, and a 2011 study found that higher BPA concentrations in the urine of the pregnant women at 16 weeks were associated with wheezing, a symptom of asthma, in their babies.

## Studies on Humans

### *Lang Study and Heart Disease*

The first large study of health effects on humans associated with bisphenol A exposure was published in September 2008 by Iain Lang and colleagues in the *Journal of the American Medical Association*. The cross-sectional study of almost 1,500 people assessed exposure to bisphenol A by looking at levels of the chemical in urine. The authors found that higher bisphenol A levels were significantly associated with heart disease, diabetes, and abnormally high levels of certain liver enzymes. An editorial in the same issue concludes:

"Based on this background information, the study by Lang et al, while preliminary with regard to these diseases in humans, should spur US regulatory agencies to follow the recent action taken by Canadian regulatory agencies, which have declared BPA a "toxic chemical" requiring aggressive action to limit human and environmental exposures. Alternatively, Congressional action could follow the precedent set with the recent passage of federal legislation designed to limit exposures to another family of compounds, phthalates, also used in plastic. Like BPA, phthalates are detectable in virtually everyone in the United States. This bill moves US policy closer to the European model, in which industry must provide data on the safety of a chemical before it can be used in products."

A later similar study performed by the same group of scientists, published in January 2010, confirmed, despite of lower concentrations of BPA in the second study sample, an associated increased risk for heart disease but not for diabetes or liver enzymes. Patients with the highest levels of BPA in their urine carried a 33% increased risk of coronary heart disease.

### *Other Studies*

Studies have associated recurrent miscarriage with BPA serum concentrations, oxidative stress and inflammation in postmenopausal women with urinary concentrations, externalizing behaviours in two-year old children, especially among female children, with mother's urinary concentrations, altered hormone levels in men and declining male sexual function with urinary concentrations. The Canadian Health Measures Survey, 2007 to 2009 published in 2010 found that teenagers carry 30 percent more 1 bisphenol A (BPA) in their bodies than older adults. The reason for this is not known. A 2010 study that analyzed BPA urinary concentrations has concluded that for people under 18 years of age BPA may negatively impact human immune function. A study done in 2010 reported the daily excretion levels of BPA among European adults in a large-scale and high-quality population-based sample, and it was shown that higher BPA daily excretion was associated with an increase in serum total testosterone concentration in men. A 2011 study found higher BPA levels in women with polycystic ovary syndrome compared to controls. Furthermore, researchers found a statistically significant positive association between male sex hormones and BPA in these women suggesting a potential role of BPA in ovarian dysfunction. A 2010 study found that people over age 18 with higher levels of BPA exposure had higher CMV antibody levels, which suggests their cell-mediated immune system may not be functioning properly.

### *Sexual Difficulties*

A 2009 study on Chinese workers in BPA factories found that workers were four times more likely to report erectile dysfunction, reduced sexual desire and overall dissatisfaction with their sex life than workers with no heightened BPA exposure. BPA workers were also seven times more likely to have ejaculation difficulties. They were also more likely to report reduced sexual function within one year of beginning employment at the factory, and the higher the exposure, the more likely they were to have sexual difficulties.

### Historical Studies

The first evidence of the estrogenicity of bisphenol A came from experiments on rats conducted in the 1930s, but it was not until 1997 that adverse effects of low-dose exposure on laboratory animals were first reported.

### Low-dose Exposure in Animals

| ***Dose (μg/kg/day)*** | ***Effects (measured in studies of mice or rats,descriptions (in quotes) are from Environmental Working Group)*** | ***Study Year*** |
|---|---|---|
| 0.025 | "Permanent changes to genital tract" | 2005 |
| 0.025 | "Changes in breast tissue that predispose cells to hormones and carcinogens" | 2005 |
| 1 | long-term adverse reproductive and carcinogenic effects | 2009 |
| 2 | "increased prostate weight 30%" | 1997 |
| 2 | "lower bodyweight, increase of anogenital distance in both genders, signs of early puberty and longer estrus." | 2002 |
| 2.4 | "Decline in testicular testosterone" | 2004 |
| 2.5 | "Breast cells predisposed to cancer" | 2007 |
| 10 | "Prostate cells more sensitive to hormones and cancer" | 2006 |
| 10 | "Decreased maternal behaviours" | 2002 |
| 30 | "Reversed the normal sex differences in brain structure and behaviour" | 2003 |
| 50 | Adverse neurological effects occur in non-human primates | 2008 |
| 50 | Disrupts ovarian development | 2009 |

The current U.S. human exposure limit set by the EPA is 50 μg/kg/day.

### Xenoestrogen

There is evidence that bisphenol A functions as a xenoestrogen by binding strongly to estrogen-related receptor γ (ERR-γ). This orphan receptor (endogenous ligand unknown) behaves as a constitutive activator of transcription. BPA seems to bind strongly to ERR-γ (dissociation constant = 5.5 nM), but not to the estrogen receptor (ER). BPA binding to ERR-γ preserves its basal constitutive activity. It can also protect it from deactivation from the selective estrogen receptor modulator 4-hydroxytamoxifen.

Different expression of ERR-γ in different parts of the body may account for variations in bisphenol A effects. For instance, ERR-γ has been found in high concentration in the placenta, explaining reports of high bisphenol accumulation in this tissue.

### Human Exposure Sources

"The problem is, BPA is also a synthetic estrogen, and plastics with BPA can break down, especially when they're washed, heated or stressed, allowing the chemical to leach into food and water and then enter the human body. That happens to nearly all of us; the CDC has found BPA in the urine of 93% of surveyed Americans over the age of 6. If you don't have BPA in your body, you're not living in the modern world." —The Perils of Plastic, TIME Magazine.

Bisphenol A has been known to be leached from the plastic lining of canned foods and polycarbonate plastics, especially those cleaned with harsh detergents or that contain acidic or high-temperature liquids. BPA is an ingredient in the internal coating of metal food and beverage cans used to protect the food from direct contact with the can. A recent Health Canada study found that the majority of canned soft drinks it tested had low, but measurable levels of bisphenol A. Furthermore, A study conducted by the University of Texas School of Public Health in 2010, found BPA in 63 of 105 samples of fresh and canned foods, foods sold in plastic packaging, and in cat and dog foods in cans and plastic packaging. This included fresh turkey, canned green beans, and canned infant formula. While most human exposure is through diet, exposure can also occur through air and through skin absorption.

A 2011 study published in *Environmental Health Perspectives*, "Food Packaging and Bisphenol A and Bis(2-Ethyhexyl) Phthalate Exposure: Findings from a Dietary Intervention," selected 20 participants based on their self-reported use of canned and packaged foods to study BPA. Participants ate their usual diets, followed by three days of consuming foods that were not canned or packaged. The study's findings include: 1) evidence of BPA in participants' urine decreased by 50% to 70% during the period of eating fresh foods; and 2), participants' reports of their food practices suggested that consumption of canned foods and beverages and restaurant meals were the most likely sources of exposure to BPA in their usual diets.

The researchers note that, even beyond these 20 participants, BPA exposure is widespread, with detectable levels in urine samples in more than an estimated 90% of the U.S. population.

Free BPA is found in high concentration in thermal paper and carbonless copy paper, which would be expected to be more available for exposure than BPA bound into resin or plastic. Popular uses of thermal paper include receipts, event and cinema tickets, labels, and airline tickets. A Swiss study found that 11 of 13 thermal printing papers contained 8 – 17 g/kg Bisphenol A (BPA). Upon dry finger contact with a thermal paper receipt, roughly 1 μg BPA (0.2 – 6 μg) was transferred to the forefinger and the middle finger. For wet or greasy fingers approximately 10 times more was transferred. Extraction of BPA from the fingers was possible up to 2 hours after exposure. While there is little concern for dermal absorption of BPA, free BPA can readily be transferred to skin and residues on hands can be ingested. Studies conducted by the CDC found bisphenol A in the urine of 95% of adults sampled in 1988–1994 and in 93% of children and adults tested in 2003–04. While the EPA considers exposures up to 50 μg/kg/day to be safe, the most sensitive animal studies show effects at much lower doses.

In 2009, a study found that drinking from polycarbonate bottles increased urinary bisphenol A levels by two thirds, from 1.2 micrograms/gram creatinine to 2 micrograms/gram creatinine. Consumer groups recommend that people wishing to lower their exposure to bisphenol A avoid canned food and polycarbonate plastic containers (which shares resin identification code 7 with many other plastics) unless the packaging indicates the plastic is bisphenol A-free. To avoid the possibility of BPA leaching into food or drink, the National Toxicology Panel recommends avoiding microwaving food in plastic containers, putting plastics in the dishwasher, or using harsh detergents. In the U.S., consumption of soda, school lunches, and meals prepared outside the home was statistically significantly associated with higher urinary BPA.

BPA is also used to form epoxy resin coating of water pipes. In older buildings, such resin coatings are used to avoid replacement of deteriorating hot and cold water pipes.

### Fetal and Early-childhood Exposures

Children may be more susceptible to BPA exposure than adults. A recent study found higher urinary concentrations in young children than in adults under typical exposure scenarios. This increased susceptibility is most likely based on their reduced capacity to eliminate xenobiotics and also their estimated higher daily exposure to BPA, adjusted for weight, compared to adults.

Infants fed with liquid formula are among the most exposed, and those fed formula from polycarbonate bottles can consume up to 13 micrograms of bisphenol A per kg of body weight per day (μg/kg/day).

In the US and Canada, BPA has been found in infant liquid formula in concentrations varying from 0.48 to 11 ng/g. BPA has been rarely found in infant powder formula (only 1 of 14). While breast milk is the optimal source of nutrition for infants, it is not always an option. The U.S. Department of Health & Human Services (HHS) states that "the benefit of a stable source of good nutrition from infant formula and food outweighs the potential risk of BPA exposure.".

A 2010 study of people in Austria, Switzerland, and Germany has suggested polycarbonate (PC) baby bottles as the most prominent role of exposure for infants, and canned food for adults and teenagers. In the United States, the growing concern over BPA exposure in infants in recent years has lead the manufacturers of plastic baby bottles to stop using BPA in their bottles. However, babies may still be exposed if they are fed with old or hand-me-down bottles bought before the companies stopped using BPA.

One often overlooked source of exposure occurs when a pregnant woman is exposed, thereby exposing the fetus. Animal studies have shown that BPA can be found in both the placenta and the amniotic fluid of pregnant mice.

A small US study in 2009, funded by the EWG, detected an average of 2.8 ng/mL BPA in the blood of 9 out of the 10 umbilical cords tested. After the baby is born, maternal exposure can continue to effect the infant through transfer of BPA to the infant via breast milk. Because of these exposures that can occur both during and after pregnancy, mothers wishing to limit their child's exposure to BPA should attempt to limit their own exposures during that time period.

While the majority of exposures have been shown to come through the diet, accidental ingestion can also be considered a source of exposure. One study conducted in Japan tested plastic baby books to look for possible leaching into saliva when babies chew on them. While the results of this study have yet to be replicated, it gives reason to question whether exposure can also occur in infants through ingestion by chewing on certain books or toys.

| **Population** | **Estimated daily bisphenol A intake, μg/kg/day. Table adapted from the National Toxicology Program Expert Panel Report.** |
|---|---|
| Infant (0–6 months)formula-fed | 1–24 |
| Infant (0–6 months)breast-fed | 0.2–1 |
| Infant (6–12 months) | 1.65–13 |
| Child (1.5–6 years) | 0.043–14.7 |
| Adult | 0.008–1.5 |

## Pharmacokinetics

There is no agreement between scientists of a physiologically-based pharmacokinetic (PBPK) BPA model for humans. The effects of BPA on an organism depend on how much free BPA is available and for how long cells are exposed to it. Glucuronidation in the liver, by conjugation with glucuronic acid to form the metabolite BPA-glucuronide (BPAG), reduces the amount of free BPA, however BPAG can be deconjugated by beta-glucuronidase, an enzyme present in high concentration in placenta and other tissues. Free BPA can also be inactivated by sulfation, a process that can also be reverted by arylsulfatase C. A 2010 vitro study has shown that placenta P-glycoprotein may efflux BPA from placenta.

The best test methods for studying BPA effects are currently under discussion with scientists sharing different opinions.

A 2010 review of 80+ biomonitoring studies concluded that the general population is internally exposed to significant amounts of unconjugated BPA (in the ng/ml blood range). Using GC/MS on 20 samples, BPA was detected in 100% of urine samples with a median of 1.25 ng/ml, and 10% of blood samples (LOD 0.5 ng/ml).

A 2009 research has found that some drugs, like naproxen, salicylic acid, carbamazepine and mefenamic acid can, *in vitro*, significantly inhibit BPA glucuronidation.

A 2010 study on rats embryos has found that genistein may enhance developmental toxicity of BPA.

## Environmental Risk

In general, studies have shown that BPA can affect growth,

reproduction and development in aquatic organisms. Among freshwater organisms, fish appear to be the most sensitive species. Evidence of endocrine-related effects in fish, aquatic invertebrates, amphibians and reptiles has been reported at environmentally relevant exposure levels lower than those required for acute toxicity. There is a widespread variation in reported values for endocrine-related effects, but many fall in the range of 1μg/L to 1 mg/L.

BPA can contaminate the environment either directly or through degradation of products containing BPA, such as ocean-borne plastic trash.

As an environmental contaminant this compound interferes with nitrogen fixation at the roots of leguminous plants associated with the bacterial symbiont *Sinorhizobium meliloti*. Despite a half-life in the soil of only 1–10 days, its ubiquity makes it an important pollutant. According to Environment Canada, "initial assessment shows that at low levels, bisphenol A can harm fish and organisms over time. Studies also indicate that it can currently be found in municipal wastewater."

A 2009 review of the biological impacts of plasticizers on wildlife published by the Royal Society with a focus on annelids (both aquatic and terrestrial), molluscs, crustaceans, insects, fish and amphibians concluded that BPA have been shown to affect reproduction in all studied animal groups, to impair development in crustaceans and amphibians and to induce genetic aberrations.

A large 2010 study of two rivers in Canada found that areas contaminated with hormone-like chemicals including bisphenol A showed females made up 85 per cent of the population of a certain fish, while females made up only 55 per cent in uncontaminated areas.

## Government and Industry Response

### *World Health Organization*

Arguing uncertainty of possible adverse health effects of low dose BPA exposure, especially on the nervous system and on behaviour, and also the differences of exposure of very young children, the WHO announced in November 2009 that it would organize an expert consultation in 2010 to assess BPA safety.

The WHO expert panel recommended no new regulations limiting or banning the use of Bisphenol-A; stating that "initiation of public health measures would be premature."

### *Australia and New Zealand*

The Australia and New Zealand Food Safety Authority (Food Standards Australia New Zealand) does not see any health risk with bisphenol A baby bottles if the manufacturer's instructions are followed. Levels of exposure are very low and do not pose a significant health risk. It added that "the move by overseas manufacturers to stop using BPA in baby bottles is a voluntary action and not the result of a specific action by regulators." It suggests the use of glass baby bottles if parents have any concerns.

### *Canada*

In April 2008, Health Canada concluded that, while adverse health effects were not expected, the margin of safety was too small for formula-fed infants and proposed classifying the chemical as "'toxic' to human health and the environment."

After the release of that assessment, Canadian Health Minister Tony Clement announced Canada's intent to ban the import, sale, and advertisement of polycarbonate baby bottles containing bisphenol A due to safety concerns, and investigate ways to reduce BPA contamination of baby formula packaged in metal cans. While the agency concluded that human exposures were less than levels believed unsafe, the margin of safety was not high enough for formula-fed infants. Around the same time, Wal-Mart announced that it was immediately ceasing sales in all its Canadian stores of food containers, water and baby bottles, sippy cups, and pacifiers containing bisphenol A, and that it would phase out baby bottles made with it in U.S. stores by early 2009. Nalgene also announced it will stop using the chemical in its products, and Toys-R-Us said it too will cease selling baby bottles made from it. Subsequent news reports showed many retailers removing polycarbonate drinking products from their shelves.

The federal government proposed declaring Bisphenol A a hazardous substance in October 2008 and has since placed it on its list of toxic substances. Health officials wrote in *Canada Gazette* that "It is concluded that bisphenol A be considered as a substance that may be entering the environment in a quantity or concentration or under conditions that constitute or may constitute a danger in Canada to human life or health." The federal ministries of health and the environment announced they would seek to restrict imports, sales and advertising of polycarbonate baby bottles containing BPA.

In its statement released on 18 October 2008, Health Canada noted that "bisphenol A exposure to newborns and infants is below levels that cause effects" and that the "general public need not be concerned".

Environment Canada listed bisphenol A as a "toxic substance" in September 2010.

### *European Union*

The updated 2008 European Union Risk Assessment Report on bisphenol A, published in June 2008, by the European Commission and European Food Safety Authority (EFSA), concluded that bisphenol A-based products, such as polycarbonate plastic and epoxy resins, are safe for consumers and the environment when used as intended.

By October 2008, after the Lang Study was published, the EFSA issued a statement concluding that the study provided no grounds to revise the current Tolerable Daily Intake (TDI) level for BPA of 0.05 mg/kg bodyweight.

A 2009 scientific study criticized the European risk assessment processes of endocrine disruptors, including BPA.

On 22 December 2009, the EU Environment ministers released a statement expressing concerns over recent studies showing adverse effects of exposure to endocrine disruptors.

In September 2010, the European Food Safety Authority (EFSA) published its latest scientific opinion, based on a "comprehensive evaluation of recent toxicity data [...] concluded that no new study could be identified, which would call for a revision of the current TDI".

The Panel noted that some studies conducted on developing animals have suggested BPA-related effects of possible toxicological relevance, in particular biochemical changes in brain, immune-modulatory effects and enhanced susceptibility to breast tumours but considered that those studies had several shortcomings so the relevance of these findings for human health could not be assessed.

On 25 November 2010, the European Union executive commission said it will ban the manufacturing by 1 March 2011 and ban the marketing and market placement of polycarbonate baby bottles containing the organic compound bisphenol A (BPA) by 1 June 2011, according to John Dalli, commissioner in charge of health and consumer policy. This is backed by a majority of EU governments. The ban was called an over-reaction by Richard Sharpe, of the Medical Research Council's Human Reproductive Sciences Unit, who said to be unaware of any convincing evidence justifying the measure and criticized it as being done on political, rather than scientific grounds.

### Denmark

In May 2009, the Danish parliament passed a resolution to ban the use of BPA in baby bottles, which has not been enacted by April 2010. In March 2010, a temporary ban was declared by the Health Minister.

### Belgium

On March 2010, senator Philippe Mahoux proposed legislation to ban BPA in food contact plastics.

### France

On 5 February 2010, the French Food Safety Agency (AFSSA) questioned the previous assessments of the health risks of BPA, especially in regard to behavioural effects observed in rat pups following exposure in utero and during the first months of life. In April 2010, the AFFSA suggested the adoption of better labels for food products containing BPA.

On 24 March 2010, French Senate unanimously approved a proposition of law to ban BPA from baby bottles. The National Assembly (Lower House) approved the text on 23 June 2010, which has been applicable law since 2 July 2010. On October 12, 2011, the French National Assembly voted a law forbidding the use of Bisphenol A in products aimed at less than 3 year old children for 2013, and 2014 for all food containers.

### Germany

On 19 September 2008, the German Federal Institute for Risk Assessment (Bundesinstitut für Risikobewertung, BfR) stated that there was no reason to change the current risk assessment for bisphenol A on the basis of the Lang Study.

In October 2009, the German environmental organization Bund für Umwelt und Naturschutz Deutschland requested a ban on BPA for children's products, especially pacifiers, and products that make contact with food. In response, some manufacturers voluntarily removed the problematic pacifiers from the market.

### Netherlands

On 6 November 2008, the Dutch Food and Consumer Product Safety Authority (VWA) stated in a newsletter that baby bottles made from polycarbonate plastic do not release measurable concentrations of bisphenol A and therefore are safe to use.

### Switzerland

In February 2009, the Swiss Federal Office for Public Health, based on reports of other health agencies, stated that the intake of bisphenol A from food represents no risk to the consumer, including newborns and infants. However, in the same statement, it advised for proper use of polycarbonate baby bottles and listed alternatives.

### Sweden

By May 2010, the Swedish Chemicals Agency asked for a BPA ban in baby bottles, but the Swedish Food Safety Authority prefers to await the expected European Food Safety Authority's updated review. The Minister of Environment said to wait for the EFSA review but not for too long. From March 2011 it is prohibited to manufacture babybottles containing bisphenol A and from July 2011 will no longer go to buy them in stores.

### UK

In December 2009, responding to a letter from a group of seven scientists that urged the UK Government to "adopt a standpoint consistent with the approach taken by other Governments who have ended the use of BPA in food contact products marketed at children", the UK Food Standards Agency reaffirmed, in January 2009, its view that "exposure of UK consumers to BPA from all sources, including food contact materials, was well below levels considered harmful".

### Turkey

As of 10 June 2011, Turkey banned the use of BPA in baby bottles and other PC items produced for babies.

### Japan

Between 1998 and 2003, the canning industry voluntarily replaced their BPA-containing epoxy resin can liners with BPA-free polyethylene terephthalate (PET) in many of their products. For other products, they switched to a different epoxy lining that yielded much less migration of BPA into food than the previously used resin. In addition, polycarbonate tableware for school lunches was replaced by BPA-free plastics. As a result of these changes, Japanese risk assessors have found that virtually no BPA is detectable in canned foods or drinks, and blood levels of BPA in people have declined dramatically (50% in one study).

### United States

#### September 2008

In September, the National Toxicology Program finalized their report on bisphenol A, finding "some concern", mid-point of a five-level scale, that infants were at risk from exposure to the chemical.

At that time, the FDA reassured consumers that current limits were safe, but convened an outside panel of experts to review the issue. The Lang study was also released that month, and David Melzer, a co-author of the study, presented the results of the study before the FDA panel.

The editorial accompanying the Lang study's publication in *JAMA* criticized the FDA's assessment of bisphenol A: "A fundamental problem is that the current ADI [acceptable daily intake] for BPA is based on experiments conducted in the early 1980s using outdated methods (only very high doses were tested) and insensitive assays. More recent findings from independent scientists were rejected by the FDA, apparently because those investigators did not follow the outdated testing guidelines for environmental chemicals, whereas studies using the outdated, insensitive assays (predominantly involving studies funded by the chemical industry) are given more weight in arriving at the conclusion that BPA is not harmful at current exposure levels."

#### March 2009

Sunoco, a producer of gasoline and chemicals, is now refusing to sell the chemical to companies for use in food and water containers for children younger than 3, saying it can't be certain of the compound's safety. Sunoco plans to require its customers to guarantee that the chemical will not be used in children's food products.

The six largest US companies that produce baby bottles decided to stop using bisphenol A in their products. Suffolk County, New York banned baby beverage containers made with bisphenol A.

On 13 March, leaders from the House and Senate proposed legislation to ban bisphenol A.

In the same month, Rochelle Tyl, author of two studies used by FDA to assert BPA safety in August 2008, said those studies didn't claim that BPA is safe because they weren't designed to cover all aspects of the chemical's effects.

### May 2009

The first US jurisdictions to pass regulations limiting or banning BPA were Minnesota and Chicago. Minnesota's regulation takes effect in 2010, "manufacturers of... children's products containing BPA may not sell them in the state after 1 Jan. 2010. The ban extends to all retailers in the state a year later." The products impacted are known as sippy cups and baby bottles. The City of Chicago adopted a similar ban shortly thereafter. Coverage of Chicago's ban in the news showed a relentless opposition by the industry. A Chicago Tribune article noted an up-hill battle while passing legislation, "[industry officials] used FDA's position on the issue when they tried to block the city's measure."

In May 2009, the Washington Post accused the manufacturers of food and beverage containers and some of their biggest customers of trying to devise a public relations and lobbying strategy to block government BPA bans. Lyndsey Layton, from the Washington Post, criticized the FDA noting that, "Despite more than 100 published studies by government scientists and university laboratories that have raised health concerns about the chemical, the Food and Drug Administration has deemed it safe largely because of two studies, both funded by a chemical industry trade group".

### June 2009

In June 2009, the FDA announced the decision to reconsider the BPA safety levels.

Connecticut was the first US state to ban bisphenol A from infant formula and baby food containers, as well from any reusable food or beverage container.

### July 2009

The California Environmental Protection Agency's Developmental and Reproductive Toxicant Identification Committee unanimously voted against placing Bisphenol A on the state's list of chemicals that are believed to cause reproductive harm. The panel, although concerned over the growing scientific research showing BPA's reproductive harm in animals, found that there was insufficient data of the effects in humans. Critics point out that the same panel failed to add second-hand smoke to the list until 2006, and only one chemical was added to the list in the last three years.

### August 2009

On 3 August, Massachusetts' Department of Public Health advised mothers to take certain actions to prevent possible health impact in children. Mothers with children up to two years old were advised to limit exposure by avoiding products that might contain BPA, such as plastic drinking bottles and other plastic materials with recycling codes of 7 or 3.

The *Milwaukee Journal Sentinel*, as part of an ongoing investigative series into BPA and its effects, revealed plans by the Society of the Plastics Industry to execute a major public relations blitz to promote BPA, including plans to attack and discredit those who report or comment negatively on the monomer and its effects.

### September 2009

On 29 September, the U.S. Environmental Protection Agency announced that it is evaluating BPA, and another five chemicals, for action plan development.

### October 2009

On 28 October, the NIH announced $30,000,000 in stimulus grants to study the health effects of BPA. This money is expected to result in many peer-reviewed publications.

### November 2009

The *Consumer Reports* magazine published an analysis of BPA content in some canned foods and beverages, where in specific cases the content of a single can of food could exceed the current FDA Cumulative Exposure Daily Intake.

### January 2010

On 15 January, the FDA expressed "some concern", the middle level in the scale of concerns, about the potential effects of BPA on the brain, behaviour, and prostate gland in fetuses, infants, and young children, and announced it was taking reasonable steps to reduce human exposure to BPA in the food supply. However, the FDA is not recommending that families change the use of infant formula or foods, as it sees the benefit of a stable source of good nutrition as outweighing the potential risk from BPA exposure.

On the same date, the U.S. Department of Health & Human Services released information to help parents to reduce children's BPA exposure.

### *February 2010*

According to *The Milwaukee Journal Sentinel*, which supports a BPA ban, after lobbyists for the chemical industry met with administration officials, the EPA delayed BPA regulation and did not include the chemical in an action plan released 30 December 2009.

Many US states are considering some sort of BPA ban.

### *March 2010*

On 29 March, the EPA declared BPA a "chemical of concern".

### *April 2010*

The 2008–2009 Annual Report of the President's Cancer Panel declared: "because of the long latency period of many cancers, the available evidence argues for a precautionary approach to these diverse chemicals, which include (...) bisphenol A"

Meanwhile, as of April 2011, General Mills has announced that it has found a BPA-free alternative can liner that apparently works even with tomatoes, a highly acidic product that has long baffled the industry in terms of finding a suitable substitute. General Mills says that with the next tomato harvest, it will begin using the BPA-free alternative in tomato products sold by its organic foods subsidiary Muir Glen. Thus far, there has been no word on whether General Mills will use BPA-free alternatives on any of its other canned products.

### *February 2011*

In August 2010, the Maine Board of Environmental Protection voted unanimously to ban the sale of baby bottles and other reusable food and beverage containers made with bisphenol A as of January 2012. In February 2011, the newly elected governor of Maine, Paul LePage, gained national attention when he spoke on a local TV news show saying he hoped to repeal the ban because, "There hasn't been any science that identifies that there is a problem" and added: "The only thing that I've heard is if you take a plastic bottle and put it in the microwave and you heat it up, it gives off a chemical similar to estrogen. So the worst case is some women may have little beards." In April, the Maine legislature passed a bill to ban the use of BPA in baby bottles, sippy cups, and other reusable food and beverage containers, effective January 1, 2012. Governor LePage refused to sign the bill.

## Polycarbonate

Polycarbonates, known by the trademarked names Lexan, Makrolon, Makroclear and others, are a particular group of thermoplastic polymers. They are easily worked, moulded, and thermoformed. Because of these properties, polycarbonates find many applications. Polycarbonates do not have a unique plastic identification code and are identified as Other, 7.

### *Structure*

Polycarbonates received their name because they are polymers containing carbonate groups (–O–(C=O)–O–). Most polycarbonates of commercial interest are derived from rigid monomers. A balance of useful features including temperature resistance, impact resistance and optical properties position polycarbonates between commodity plastics and engineering plastics.

### *Production*

The main polycarbonate material is produced by the reaction of bisphenol A and phosgene $COCl_2$. The overall reaction can be written as follows:

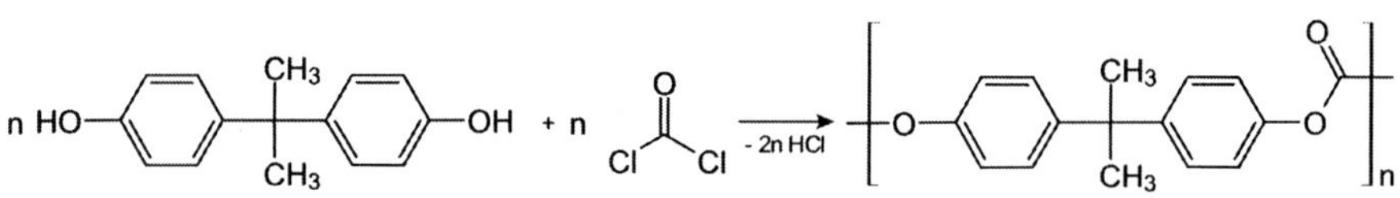


The first step of the synthesis involves treatment of bisphenol A with sodium hydroxide, which deprotonates the hydroxyl groups of the bisphenol A.

$$(HOC_6H_4)_2CMe_2 + 2\ NaOH \rightarrow (NaOC_6H_4)_2CMe_2 + 2\ H_2O$$

The diphenoxide ($(NaOC_6H_4)_2CMe_2$) reacts with phosgene to give a chloroformate, which subsequently is attacked by another phenoxide. The net reaction from the diphenoxide is:

$$(NaOC_6H_4)_2CMe_2 + COCl_2 \rightarrow 1/n\ [OC(OC_6H_4)_2CMe_2]_n + 2\ NaCl$$

In this way, approximately one billion kilograms of polycarbonate is produced annually. Many other diols have been tested in place of bisphenol A, e.g. 1,1-bis(4-hydroxyphenyl)cyclohexane and dihydroxybenzophenone including some, e.g. tetramethylcyclobutanediol, that are unlikely endocrine disruptors.

An alternative route to polycarbonates entails transesterification from BPA and diphenyl carbonate:

$(HOC_6H_4)_2CMe_2 + (C_6H_5O)_2CO \rightarrow 1/n\ [OC(OC_6H_4)_2CMe_2]_n + 2\ C_6H_5OH$

The diphenyl carbonate was derived in part from carbon monoxide, this route being greener than the phosgene method.

### Properties and Processing

Polycarbonate is a very durable material. Although it has high impact-resistance, it has low scratch-resistance and so a hard coating is applied to polycarbonate eyewear lenses and polycarbonate exterior automotive components. The characteristics of polycarbonate are quite like those of polymethyl methacrylate (PMMA, acrylic), but polycarbonate is stronger, usable in a wider temperature range but more expensive. This polymer is highly transparent to visible light and has better light transmission characteristics than many kinds of glass.

Polycarbonate has a glass transition temperature of about 150 °C (302 °F), so it softens gradually above this point and flows above about 300 °C (572 °F). Tools must be held at high temperatures, generally above 80 °C (176 °F) to make strain- and stress-free products. Low molecular mass grades are easier to mould than higher grades, but their strength is lower as a result. The toughest grades have the highest molecular mass, but are much more difficult to process.

Unlike most thermoplastics, polycarbonate can undergo large plastic deformations without cracking or breaking. As a result, it can be processed and formed at room temperature using sheet metal techniques, such as forming bends on a brake. Even for sharp angle bends with a tight radius, no heating is generally necessary. This makes it valuable in prototyping applications where transparent or electrically non-conductive parts are needed, which cannot be made from sheet metal. Note that PMMA/Plexiglas, which is similar in appearance to polycarbonate, is brittle and cannot be bent at room temperature.

Main transformation techniques for polycarbonate resins:

- extrusion into tubes, rods and other profiles
- extrusion with cylinders into sheets (0.5–15 mm (0.020–0.59 in)) and films (below 1 mm (0.039 in)), which can be used directly or manufactured into other shapes using thermoforming or secondary fabrication techniques, such as bending, drilling, routing, laser cutting etc.
- injection molding into ready articles

### Applications

#### Electronic Components

Polycarbonate is mainly used for electronic applications that capitalize on its collective safety features. Being a good electrical insulator and having heat resistant and flame retardant properties, it is used in various products associated with electrical and telecommunications hardware. It also serves as dielectric in high stability capacitors.

#### Construction Materials

The second largest consumer of polycarbonates is the construction industry, e.g. for domelights, flat or curved glazing, and sound walls.

#### Data Storage

A major application of polycarbonate is the production of Compact Discs, DVDs, and Blu-ray Discs. These discs are produced by injection molding polycarbonate into a mold cavity that has on one side a metal stamper containing a negative image of the disc data, while the other mold side is a mirrored surface. Typical products of sheet/film production include applications in advertisement (signs, displays, poster protection).

#### Automotive, Aircraft, and Security Components

In the automotive industry, injection-moulded polycarbonate can produce very smooth surfaces that make it well-suited for direct (without the need for a basecoat) metalised parts such as decorative bezels and optical reflectors. Its uniform mould shrinkage results in parts with greater accuracy than those made of polypropylene. However, due to its susceptibility to environmental stress cracking, its use is limited to low-stress applications. It can be laminated to make bullet-proof "glass", although "bullet-resistant" is more accurate for the thinner windows, such as are used in bullet-resistant windows in automobiles. The thicker barriers of transparent plastic used in teller's windows and barriers in banks are also polycarbonate.

So-called "theft-proof" large plastic packaging for smaller items, which cannot be opened by hand, is uniformly made from polycarbonate.

The cockpit canopy of the F-22 Raptor jet fighter is made from a piece of high optical quality polycarbonate, and is the largest piece of its type formed in the world.

### Niche Applications

Polycarbonate, being a versatile material with attractive processing and physical properties, has attracted myriad smaller applications. The use of injection molded drinking bottles and glasses and food containers has stirred serious controversy.

Many kinds of lenses are manufactured from polycarbonate, including durable, lightweight luggage, automotive headlamp lenses, lighting lenses, sunglass/eyeglass, lenses, and safety glasses. Other miscellaneous items: MP3/digital audio player cases, ocarinas, computer cases, riot shields, visors, instrument panels, and blender jars. Many toys and hobby items are made from polycarbonate parts, e.g. fins, gyro mounts, and flybar locks for use with radio-controlled helicopters.

For use in applications exposed to weathering or UV-radiation, a special surface treatment is needed. This either can be a coating (e.g. for improved abrasion resistance), or a coextrusion for enhanced weathering resistance.

Polycarbonate is also used as a printing substrate for nameplate and other forms of industrial grade under printed products. The polycarbonate provides a barrier to wear, the elements, and fading.

### Medical Applications

Some polycarbonate grades are used in medical applications and comply with both ISO 10993-1 and USP Class VI standards (occasionally referred to as PC-ISO). Class VI is the most stringent of the six USP ratings. These grades can be sterilized using steam at 120 °C, gamma radiation, or by the ethylene oxide (EtO) method. However, scientific research indicates possible problems with biocompatibility. Dow Chemical strictly limits all its plastics with regard to medical applications.

### *Potential Hazards in Food Contact Applications*

The use of polycarbonate containers for the purpose of food storage is controversial. The basis of this controversy is their hydrolysis (degradation by water, often referred to as leaching) releases bisphenol A:

$$1/n\ [OC(OC_6H_4)_2CMe_2]_n + H_2O \rightarrow (HOC_6H_4)_2CMe_2 + CO_2$$

More than 100 studies have explored the bioactivity of bisphenol A derived from polycarbonates. Bisphenol A appeared to be released from polycarbonate animal cages into water at room temperature and it may have been responsible for enlargement of the reproductive organs of female mice. However, the animal cages used in the research were fabricated from industrial grade polycarbonate, rather than FDA food grade polycarbonate.

An analysis of the literature on bisphenol A leachate low-dose effects by vom Saal and Hughes published in August 2005 seems to have found a suggestive correlation between the source of funding and the conclusion drawn. Industry funded studies tend to find no significant effects whereas government funded studies tend to find significant effects.

Sodium hypochlorite bleach and other alkali cleaners catalyze the release of the bisphenol A from polycarbonate containers. A chemical compatibility chart shows that polycarbonate is incompatible with ammonia and acetone because it dissolves in their presence. Alcohol is one recommended organic solvent for cleaning grease and oils from polycarbonate.

## Phosgene

Phosgene is the chemical compound with the formula $COCl_2$. This colorless gas gained infamy as a chemical weapon during World War I. It is also a valued industrial reagent and building block in synthesis of pharmaceuticals and other organic compounds. In low concentrations, its odor resembles freshly cut hay or grass.

In addition to its industrial production, small amounts occur naturally from the breakdown and the combustion of organochlorine compounds, such as those used in refrigeration systems. The name, sounding similar to "phosphine", does not mean it has any phosphorus. The chemical was named by combining the Greek words 'phos' (meaning light) and genesis (birth).

### *Structure and Basic Properties*

Phosgene is a planar molecule as predicted by VSEPR theory. The C=O distance is 1.18 Å, the C—Cl distance is 1.74 Å and the Cl—C—Cl angle is 111.8°. It is one of the simplest acid chlorides, being formally derived from carbonic acid.

### *Production*

Industrially, phosgene is produced by passing purified carbon monoxide and chlorine gas through a bed of porous activated carbon, which serves as a catalyst:

$$CO + Cl_2 \rightarrow COCl_2 \ (\Delta H_{rxn} = -107.6 kJ/mol)$$

The reaction is exothermic, therefore the reactor must be cooled. Typically, the reaction is conducted between 50 and 150 °C. Above 200 °C, phosgene reverts to carbon monoxide and chlorine, $K_{eq}$ (300K) = 0.05. Approximately 5000 tonnes were produced in 1989.

Because of safety issues, phosgene is almost always produced and consumed within the same plant and extraordinary measures are made to contain this toxic gas. It is listed on schedule 3 of the Chemical Weapons Convention: All production sites manufacturing more than 30 tonnes per year must be declared to the OPCW. Although less dangerous than many other chemical weapons, such as sarin, phosgene is still regarded as a viable chemical warfare agent because it is so easy to manufacture when compared to the production requirements of more technically advanced chemical weapons such as the first-generation nerve agent tabun.

### *Adventitious Occurrence*

Upon ultraviolet (UV) radiation in the presence of oxygen, chloroform slowly converts into phosgene via a radical reaction. To suppress this photodegradation, chloroform is often stored in brown-tinted glass containers. Chlorinated compounds used to remove oil from metals, such as automotive brake cleaners, are converted to phosgene by the UV rays of arc welding processes.

Phosgene may also be produced during testing for leaks of older-style refrigerant gasses. Chloromethanes (R12, R22 and others) were formerly leak-tested in situ by employing a small gas torch (propane, butane or propylene gas) with a sniffer tube and a copper reaction plate in the flame nozzle of the torch. If any refrigerant gas was leaking from a pipe or joint, the gas would be sucked into the flame via the sniffer tube and would cause a colour change of the gas flame to a bright greenish blue. In the process, phosgene gas would be created due to the thermal reaction.

No valid statistics are available, but anecdotal reports suggest that numerous refrigeration technicians suffered the effects of phosgene poisoning due to their ignorance of the toxicity of phosgene, produced during such leak testing.

Electronic sensing of refrigerant gases phased out the use of flame testing for leaks in the 1980s. Similarly, phosgene poisoning is a consideration for people fighting fires that are occurring in the vicinity of freon refrigeration equipment, smoking in the vicinity of a freon leak, or fighting fires using halon or halotron.

### Uses

The great majority of phosgene is used in the production of isocyanates, the most important being toluene diisocyanate (TDI) and methylene diphenyl diisocyanate (MDI). These isocyanates are precursors to polyurethanes. Significant amounts are also used in the production of polycarbonates via its reaction with bisphenol A. Polycarbonates are an important class of engineering thermoplastic found, for example, in lenses in eye glasses.

### Organic Synthesis

Although phosgene still finds use in organic synthesis, a variety of substitutes have been developed, notably trichloromethyl chloroformate ("diphosgene"), which is a liquid at room temperature, and bis(trichloromethyl) carbonate ("triphosgene"), a crystalline substance. The following are the three most useful reactions involving phosgene.

### Synthesis of Carbonates

Diols react with phosgene to give either linear or cyclic carbonates (R = H, alkyl, aryl):

$$HOCR_2\text{-}X\text{-}CR_2OH + COCl_2 \rightarrow 1/n\ [OCR_2\text{-}X\text{-}CR_2OC(O)\text{-}]_n + 2\ HCl$$

### Synthesis of Isocyanates

The synthesis of isocyanates from amines illustrates the electrophilic character of this reagent and its use in introducing the equivalent of "$CO^{2+}$" (R = alkyl, aryl):

$$RNH_2 + COCl_2 \rightarrow RN{=}C{=}O + 2\ HCl$$

Such reactions are conducted in the presence of a base such as pyridine that absorbs the hydrogen chloride.

### Synthesis of Acid Chlorides

It is also used to produce acid chlorides and carbon dioxide from carboxylic acids:

$$RCO_2H + COCl_2 \rightarrow RC(O)Cl + HCl + CO_2$$

Such acid chlorides react with amines and alcohols to give, respectively, amides and esters, which are commonly used intermediates. Thionyl chloride is more commonly and more safely employed for this application. A specific application for phosgene is the production of chloroformic esters:

$$ROH + COCl_2 \rightarrow ROC(O)Cl + HCl$$

### Inorganic Chemistry

Although it is somewhat hydrophobic, phosgene reacts with water to release hydrogen chloride and carbon dioxide:

$$COCl_2 + H_2O \rightarrow CO_2 + 2\ HCl$$

Analogously, with ammonia, one obtains urea:

$$COCl_2 + 4\ NH_3 \rightarrow CO(NH_2)_2 + 2\ NH_4Cl$$

Halide exchange with nitrogen trifluoride and aluminium tribromide gives $COF_2$ and $COBr_2$, respectively.

### History

Phosgene was synthesized by the British chemist John Davy (1790–1868) in 1812 by exposing a mixture of carbon monoxide and chlorine to sunlight. He named it “phosgene” in reference of the use of light to promote the reaction; from Greek, *phos* (light) and *gene* (born). It gradually became important in the chemical industry as the 19th century progressed, particularly in dye manufacturing.

### Chemical Warfare

Following the extensive use of phosgene gas in combat during World War I, it was stockpiled by various countries as part of their secret chemical weapons programs.

Phosgene was then only frequently used by the Imperial Japanese Army against the Chinese during the Second Sino-Japanese War. Gas weapons, such as phosgene, were produced by Unit 731 and authorized by specific orders given by Hirohito (Emperor Showa) himself, transmitted by the chief of staff of the army. For example, the Emperor authorized the use of toxic gas on 375 separate occasions during the battle of Wuhan from August to October 1938.

### *Safety*

Phosgene is an insidious poison as the odor may not be noticed and symptoms may be slow to appear. Phosgene can be detected at 0.4 ppm, which is four times the Threshold Limit Value. Its high toxicity arises from the action of the phosgene on the proteins in the pulmonary alveoli, which are the site of gas exchange: their damage disrupts the blood-air barrier, causing suffocation. It reacts with the amines of the proteins, causing crosslinking via formation of urea-like linkages, in accord with the reactions. Phosgene detection badges are worn by those at risk of exposure.

Sodium bicarbonate may be used to neutralise liquid spills of phosgene. Gaseous spills may be mitigated with ammonia.

## Solvent

A solvent (from the Latin *solvô*, "I loosen, untie, I solve") is a liquid, solid, or gas that dissolves another solid, liquid, or gaseous solute, resulting in a solution that is soluble in a certain volume of solvent at a specified temperature. Common uses for organic solvents are in dry cleaning (e.g., tetrachloroethylene), as a paint thinner (e.g., toluene, turpentine), as nail polish removers and glue solvents (acetone, methyl acetate, ethyl acetate), in spot removers (e.g., hexane, petrol ether), in detergents (citrus terpenes), in perfumes (ethanol), and nail polish, (part of how it stays on nails) and in chemical synthesis. The use of inorganic solvents (other than water) is typically limited to research chemistry and some technological processes.

In 2005, the worldwide market for solvents had a total volume of around 17.9 million tons, which led to a turnover of about €8,000,000,000.

### *Solutions and Solvation*

When one substance is dissolved into another, a solution is formed. This is opposed to the situation when the compounds are insoluble and one of them precipitate like sand in water. In solution, all of the ingredients are uniformly distributed at a molecular level and no residue remains. The mixing is referred to as miscibility, whereas the ability to dissolve one compound into another is known as solubility.

However, in addition to mixing, both substances in the solution interact with each other. When something is dissolved, molecules of the solvent arrange themselves around molecules of the solute. Heat is involved and entropy is increased making the solution more thermodynamically stable than the solute alone. This arrangement is mediated by the respective chemical properties of the solvent and solute, such as hydrogen bonding, dipole moment and polarizability.

### Solvent Classifications

Solvents can be broadly classified into two categories: *polar* and *non-polar*. Generally, the dielectric constant of the solvent provides a rough measure of a solvent's polarity. The strong polarity of water is indicated, at 20 °C, by a dielectric constant of 80.10;. Solvents with a dielectric constant of less than 15 are generally considered to be nonpolar. Technically, the dielectric constant measures the solvent's ability to reduce the field strength of the electric field surrounding a charged particle immersed in it. This reduction is then compared to the field strength of the charged particle in a vacuum. In laymen's terms, dielectric constant of a solvent can be thought of as its ability to reduce the solute's internal charge.

### Other Polarity Scales

Dielectric constants are not the only measure of polarity. Because solvents are used by chemists to carry out chemical reactions or observe chemical and biological phenomena, more specific measures of polarity are required.

*The Grunwald Winstein mY scale* measures polarity in terms of solvent influence on buildup of positive charge of a solute during a chemical reaction.

*Kosower's Z scale* measures polarity in terms of the influence of the solvent on uv absorption maxima of a salt, usually pyridinium iodide or the pyridinium zwitterion.

*Donor number and donor acceptor scale* measures polarity in terms of how a solvent interacts with specific substances, like a strong Lewis acid or a strong Lewis base.

The polarity, dipole moment, polarizability and hydrogen bonding of a solvent determines what type of compounds it is able to dissolve and with what other solvents or liquid compounds it is miscible. As a rule of thumb, polar solvents dissolve polar compounds best and non-polar solvents dissolve non-polar compounds best: "like dissolves like". Strongly polar compounds like sugars (e.g., sucrose) or ionic compounds, like inorganic salts (e.g., table salt) dissolve only in very polar solvents like water, while strongly non-polar compounds like oils or waxes dissolve only in very non-polar organic solvents like hexane. Similarly, water and hexane (or vinegar and vegetable oil) are not miscible with each other and will quickly separate into two layers even after being shaken well.

### *Polar Protic and Polar Aprotic*

Solvents with a relative static permittivity greater than 15 can be further divided into protic and aprotic. Protic solvents solvate anions (negatively charged solutes) strongly via hydrogen bonding. Water is a protic solvent. Aprotic solvents such as acetone or dichloromethane tend to have large dipole moments (separation of partial positive and partial negative charges within the same molecule) and solvate positively charged species via their negative dipole. In chemical reactions the use of polar protic solvents favors the $S_N1$ reaction mechanism, while polar aprotic solvents favour the $S_N2$ reaction mechanism.

## Physical Properties of Common Solvents

### *Properties Table of Common Solvents*

The solvents are grouped into non-polar, polar aprotic, and polar protic solvents and ordered by increasing polarity. The polarity is given as the dielectric constant. The properties of solvents that exceed those of water are bolded.

| *Solvent* | *Chemical formula* | *Boiling point* | *Dielectric constant* | *Density* | *Dipole moment* |
|---|---|---|---|---|---|
| | *Non-polar solvents* | | | | |
| Pentane | $CH_3$-$CH_2$-$CH_2$-$CH_2$-$CH_3$ | 36 °C | 1.84 | 0.626 g/ml | 0.00 D |
| Cyclopentane | $C_5H_{10}$ | 40 °C | 1.97 | 0.751 g/ml | 0.00 D |
| Hexane | $CH_3$-$CH_2$-$CH_2$-$CH_2$-$CH_2$-$CH_3$ | 69 °C | 1.88 | 0.655 g/ml | 0.00 D |
| Cyclohexane | $C_6H_{12}$ | 81 °C | 2.02 | 0.779 g/ml | 0.00 D |
| Benzene | $C_6H_6$ | 80 °C | 2.3 | 0.879 g/ml | 0.00 D |
| Toluene | $C_6H_5$-$CH_3$ | 111 °C | 2.38 | 0.867 g/ml | 0.36 D |
| 1,4-Dioxane | /-$CH_2$-$CH_2$-O-$CH_2$-$CH_2$-O-\ | 101 °C | 2.3 | 1.033 g/ml | 0.45 D |

| | | | | | |
|---|---|---|---|---|---|
| Chloroform | $CHCl_3$ | 61 °C | 4.81 | 1.498 g/ml | 1.04 D |
| Diethyl ether | $CH_3CH_2$-O-$CH_2$-$CH_3$ | 35 °C | 4.3 | 0.713 g/ml | 1.15 D |
| | ***Polar aprotic solvents*** | | | | |
| Dichloromethane (DCM) | $CH_2Cl_2$ | 40 °C | 9.1 | 1.3266 g/ml | 1.60 D |
| Tetrahydrofuran (THF) | /-$CH_2$-$CH_2$-O-$CH_2$-$CH_2$-\ | 66 °C | 7.5 | 0.886 g/ml | 1.75 D |
| Ethyl acetate | $CH_3$-C(=O)-O-$CH_2$-$CH_3$ | 77 °C | 6.02 | 0.894 g/ml | 1.78 D |
| Acetone | $CH_3$-C(=O)-$CH_3$ | 56 °C | 21 | 0.786 g/ml | 2.88 D |
| Dimethylformamide (DMF) | H-C(=O)N$(CH_3)_2$ | 153 °C | 38 | 0.944 g/ml | 3.82 D |
| Acetonitrile (MeCN) | $CH_3$-Ca”N | 82 °C | 37.5 | 0.786 g/ml | 3.92 D |
| Dimethyl sulfoxide (DMSO) | $CH_3$-S(=O)-$CH_3$ | 189 °C | 46.7 | 1.092 g/ml | 3.96 D |
| | ***Polar protic solvents*** | | | | |
| Formic acid | H-C(=O)OH | 101 °C | 58 | 1.21 g/ml | 1.41 D |
| *n*-Butanol | $CH_3$-$CH_2$-$CH_2$-$CH_2$-OH | 118 °C | 18 | 0.810 g/ml | 1.63 D |
| Isopropanol (IPA) | $CH_3$-CH(-OH)-$CH_3$ | 82 °C | 18 | 0.785 g/ml | 1.66 D |
| *n*-Propanol | $CH_3$-$CH_2$-$CH_2$-OH | 97 °C | 20 | 0.803 g/ml | 1.68 D |
| Ethanol | $CH_3$-$CH_2$-OH | 79 °C | 24.55 | 0.789 g/ml | 1.69 D |
| Methanol | $CH_3$-OH | 65 °C | 33 | 0.791 g/ml | 1.70 D |
| Acetic acid | $CH_3$-C(=O)OH | 118 °C | 6.2 | 1.049 g/ml | 1.74 D |
| Water | H-O-H | 100 °C | 80 | 1.000 g/ml | 1.85 D |

There's another powerful way to look at these same solvents. By knowing their Hansen solubility parameter values (HSPiP), which are based on δD=dispersion bonds, δP=polar bonds and δH=hydrogen bonds, you know important things about their inter-molecular interactions with other solvents and also with polymers, pigments, nanoparticles etc. so you can do two things. First, you can create rational formulations knowing, for example, that there is a good HSP match between a solvent and a polymer. Second, you can make rational substitutions for "good" solvents (they dissolve things well) that are "bad" (for the environment, for health, for cost etc.). The following table shows that the intuitions from "non-polar", "polar aprotic" and "polar protic" are put numerically - the "polar" molecules have higher levels of δP and the protic solvents have higher levels of δH. Because numerical values are used, comparisons can be made rationally by comparing numbers. So acetonitrile is much more polar than acetone but slightly less hydrogen bonding.

| ***Solvent*** | ***Chemical formula*** | ***dD Dispersion*** | ***dP Polar*** | ***dH Hydrogen bonding*** |
|---|---|---|---|---|
| | ***Non-polar solvents*** | | | |
| Hexane | $CH_3$-$CH_2$-$CH_2$-$CH_2$-$CH_2$-$CH_3$ | 14.9 | 0.0 | 0.0 |
| Benzene | $C_6H_6$ | 18.4 | 0.0 | 2.0 |
| Toluene | $C_6H_5$-$CH_3$ | 18.0 | 1.4 | 2.0 |
| Diethyl ether | $CH_3CH_2$-O-$CH_2$-$CH_3$ | 14.5 | 2.9 | 4.6 |
| Chloroform | $CHCl_3$ | 17.8 | 3.1 | 5.7 |
| 1,4-Dioxane | /-$CH_2$-$CH_2$-O-$CH_2$-$CH_2$-O-\ | 17.5 | 1.8 | 9.0 |
| | ***Polar aprotic solvents*** | | | |
| Ethyl acetate | $CH_3$-C(=O)-O-$CH_2$-$CH_3$ | 15.8 | 5.3 | 7.2 |
| Tetrahydrofuran (THF) | /-$CH_2$-$CH_2$-O-<br>$CH_2$-$CH_2$-\ | 16.8 | 5.7 | 8.0 |
| Dichloromethane | $CH_2Cl_2$ | 17.0 | 7.3 | 7.1 |
| Acetone | $CH_3$-C(=O)-$CH_3$ | 15.5 | 10.4 | 7.0 |
| Acetonitrile (MeCN) | $CH_3$-Ca"N | 15.3 | 18.0 | 6.1 |
| Dimethylformamide (DMF) | H-C(=O)N$(CH_3)_2$ | 17.4 | 13.7 | 11.3 |
| Dimethyl sulfoxide (DMSO) | $CH_3$-S(=O)-$CH_3$ | 18.4 | 16.4 | 10.2 |
| Polar protic solvents | | | | |
| Acetic acid | $CH_3$-C(=O)OH | 14.5 | 8.0 | 13.5 |
| *n*-Butanol | $CH_3$-$CH_2$-$CH_2$-$CH_2$-OH | 16.0 | 5.7 | 15.8 |
| Isopropanol | $CH_3$-CH(-OH)-$CH_3$ | 15.8 | 6.1 | 16.4 |
| *n*-Propanol | $CH_3$-$CH_2$-$CH_2$-OH | 16.0 | 6.8 | 17.4 |
| Ethanol | $CH_3$-$CH_2$-OH | 15.8 | 8.8 | 19.4 |
| Methanol | $CH_3$-OH | 14.7 | 12.3 | 22.3 |
| Formic acid | H-C(=O)OH | 14.6 | 10.0 | 14.0 |
| Water | H-O-H | 15.5 | 16.0 | 42.3 |

Consider a simple example of rational substitution. Suppose for environmental reasons we needed to replace the chlorinated solvent, chloroform, with a solvent (blend) of equal solvency using a mixture of two non-chlorinated solvents from this table. Via trial-and-error, a spreadsheet or some software such as HSPiP we find that a 50:50 mix of toluene and 1,4 dioxane is a close match. The dD of the mixture is the average of 18.0 and 17.5 = 17.8. The dP of the mixture is the average of 1.4 and 1.8 = 1.6 and the dH of the mixture is the average of 2.0 and 9.0 = 5.5. So the mixture is 17.8, 1.6, 5.5 compared to Chloroform at 17.8, 3.1, 5.7. Because Toluene itself has many health issues, other mixtures of solvents can be found using a full Hansen solubility parameter dataset.

### Boiling Point

| *Solvent* | *Boiling point (°C)* |
|---|---|
| ethylene dichloride | 83.48 |
| pyridine | 115.25 |
| methyl isobutyl ketone | 116.5 |
| methylene chloride | 39.75 |
| isooctane | 99.24 |
| carbon disulfide | 46.3 |
| carbon tetrachloride | 76.75 |
| o-xylene | 144.42 |

An important property of solvents is boiling point. This also determines the speed of evaporation. Small amounts of low-boiling solvents like diethyl ether, dichloromethane, or acetone will evaporate in seconds at room temperature, while high-boiling solvents like water or dimethyl sulfoxide need higher temperatures, an air flow, or the application of vacuum for fast evaporation.

- Low boilers: boiling temperature below 100 °C (boiling point of water)
- Medium boilers: between 100 °C and 150 °C
- High boilers: above 150 °C

### Density

Most organic solvents have a lower density than water, which means they are lighter and will form a separate layer on top of water. An important exception: most of the halogenated solvents like dichloromethane or chloroform will sink to the bottom of a container, leaving water as the top layer. This is important to remember when partitioning compounds between solvents and water in a separatory funnel during chemical syntheses. Often, specific gravity is cited in place of density. Specific gravity is defined as the density of the solvent divided by the density of water at the same temperature. As such, specific gravity is a unitless value. It readily communicates whether a water-insoluble solvent will float (SG < 1.0) or sink (SG > 1.0) when mixed with water.

| ***Solvent*** | ***Specific gravity*** |
|---|---|
| Pentane | 0.626 |
| Petroleum ether | 0.656 |
| Hexane | 0.659 |
| Heptane | 0.684 |
| Diethyl amine | 0.707 |
| Diethyl ether | 0.713 |
| Triethyl amine | 0.728 |
| Tert-butyl methyl ether | 0.741 |
| Cyclohexane | 0.779 |
| Tert-butyl alcohol | 0.781 |
| Isopropanol | 0.785 |
| Acetonitrile | 0.786 |
| Ethanol | 0.789 |
| Acetone | 0.790 |
| Methanol | 0.791 |
| Methyl isobutyl ketone | 0.798 |
| Isobutyl alcohol | 0.802 |
| 1-Propanol | 0.803 |
| Methyl ethyl ketone | 0.805 |
| 2-Butanol | 0.808 |
| Isoamyl alcohol | 0.809 |
| 1-Butanol | 0.810 |
| Diethyl ketone | 0.814 |
| 1-Octanol | 0.826 |
| p-Xylene | 0.861 |
| m-Xylene | 0.864 |
| Toluene | 0.867 |
| Dimethoxyethane | 0.868 |

| | |
|---|---|
| Benzene | 0.879 |
| Butyl acetate | 0.882 |
| 1-Chlorobutane | 0.886 |
| Tetrahydrofuran | 0.889 |
| Ethyl acetate | 0.895 |
| o-Xylene | 0.897 |
| Hexamethylphosphorus triamide | 0.898 |
| 2-Ethoxyethyl ether | 0.909 |
| N,N-Dimethylacetamide | 0.937 |
| Diethylene glycol dimethyl ether | 0.943- |
| N,N-Dimethylformamide | 0.944 |
| 2-Methoxyethanol | 0.965 |
| Pyridine | 0.982 |
| Propanoic acid | 0.993 |
| Water | 1.000 |
| 2-Methoxyethyl acetate | 1.009 |
| Benzonitrile | 1.01 |
| 1-Methyl-2-pyrrolidinone | 1.028 |
| Hexamethylphosphoramide | 1.03 |
| 1,4-Dioxane | 1.033 |
| Acetic acid | 1.049 |
| Acetic anhydride | 1.08 |
| Dimethyl sulfoxide | 1.092 |
| Chlorobenzene | 1.1066 |
| Deuterium oxide | 1.107 |
| Ethylene glycol | 1.115 |
| Diethylene glycol | 1.118 |
| Propylene carbonate | 1.21 |
| Formic acid | 1.22 |
| 1,2-Dichloroethane | 1.245 |
| Glycerin | 1.261 |
| Carbon disulfide | 1.263 |
| 1,2-Dichlorobenzene | 1.306 |

| | |
|---|---|
| Methylene chloride | 1.326 |
| Nitromethane | 1.382 |
| 2,2,2-Trifluoroethanol | 1.393 |
| Chloroform | 1.498 |
| 1,1,2-Trichlorotrifluoroethane | 1.575 |
| Carbon tetrachloride | 1.594 |
| Tetrachloroethylene | 1.623 |

## Health and Safety

### *Fire*

Most organic solvents are flammable or highly flammable, depending on their volatility. Exceptions are some chlorinated solvents like dichloromethane and chloroform. Mixtures of solvent vapours and air can explode. Solvent vapours are heavier than air; they will sink to the bottom and can travel large distances nearly undiluted. Solvent vapours can also be found in supposedly empty drums and cans, posing a flash fire hazard; hence empty containers of volatile solvents should be stored open and upside down.

Both diethyl ether and carbon disulfide have exceptionally low autoignition temperatures which increase greatly the fire risk associated with these solvents. The autoignition temperature of carbon disulfide is below 100°C (212°F), so objects such as steam pipes, light bulbs, hotplates and recently extinguished bunsen burners are able to ignite its vapours.

### *Explosive Peroxide Formation*

Ethers like diethyl ether and tetrahydrofuran (THF) can form highly explosive organic peroxides upon exposure to oxygen and light, THF is normally more able to form such peroxides than diethyl ether. One of the most susceptible solvents is diisopropyl ether.

The heteroatom (oxygen) stabilizes the formation of a free radical which is formed by the abstraction of a hydrogen atom by another free radical. The carbon centred free radical thus formed is able to react with an oxygen molecule to form a peroxide compound. A range of tests can be used to detect the presence of a peroxide in an ether; one is to use a combination of iron sulfate and potassium thiocyanate. The peroxide is able to oxidize the $Fe^{2+}$ ion to an $Fe^{3+}$ ion which then form a deep red coordination complex with the thiocyanate. In extreme cases the peroxides can form crystalline solids within the vessel of the ether.

Unless the desiccant used can destroy the peroxides, they will concentrate during distillation due to their higher boiling point. When sufficient peroxides have formed, they can form a crystalline and shock sensitive solid precipitate. When this solid is formed at the mouth of the bottle, turning the cap may provide sufficient energy for the peroxide to detonate. Peroxide formation is not a significant problem when solvents are used up quickly; they are more of a problem for laboratories which take years to finish a single bottle. Ethers have to be stored in the dark in closed canisters in the presence of stabilizers like butylated hydroxytoluene (BHT) or over sodium hydroxide.

Peroxides may be removed by washing with acidic iron(II) sulfate, filtering through alumina, or distilling from sodium/benzophenone. Alumina does not destroy the peroxides; it merely traps them. The advantage of using sodium/benzophenone is that moisture and oxygen is removed as well.

### Health Effects

General health hazards associated with solvent exposure include toxicity to the nervous system, reproductive damage, liver and kidney damage, respiratory impairment, cancer, and dermatitis.

Many solvents can lead to a sudden loss of consciousness if inhaled in large amounts. Solvents like diethyl ether and chloroform have been used in medicine as anesthetics, sedatives, and hypnotics for a long time. Ethanol (grain alcohol) is a widely used and abused psychoactive drug. Diethyl ether, chloroform, and many other solvents (e.g., from gasoline or glues) are used recreationally in glue sniffing, often with harmful long term health effects like neurotoxicity or cancer. Methanol can cause permanent blindness and death. It is also dangerous because it burns with an invisible flame.

It is interesting to note that ethanol has a synergistic effect when taken in combination with many solvents. For instance a combination of toluene/benzene and ethanol causes greater nausea/vomiting than either substance alone.

Some solvents including chloroform and benzene (an ingredient of gasoline) are carcinogenic. Many others can damage internal organs like the liver, the kidneys, or the brain.

Chronic exposure to organic solvents in the work environment can produce a range of adverse neuropsychiatric effects and, occupational exposure to organic solvents has been associated with higher numbers of painters regularly exposed to high levels of organic solvents suffering from alcoholism.

### General Precautions

- Avoid being exposed to solvent vapours by working in a fume hood, or with local exhaust ventilation (LEV), or in a well ventilated area
- Keep the storage containers tightly closed
- Never use open flames near flammable solvents; use electrical heating instead
- Never flush flammable solvents down the drain; read safety data sheets for proper disposal information
- Avoid the inhalation of solvent vapours
- Avoid contact of the solvent with the skin — many solvents are easily absorbed through the skin. They also tend to dry the skin and may cause sores and wounds.

### Environmental Contamination

A major pathway to induce health effects arises from spills or leaks of solvents that reach the underlying soil. Since solvents readily migrate substantial distances, the creation of widespread soil contamination is not uncommon; there may be about 5000 sites worldwide that have major subsurface solvent contamination; this is particularly a health risk if aquifers are affected.

## Cyclohexane

Cyclohexane is a cycloalkane with the molecular formula $C_6H_{12}$. Cyclohexane is used as a nonpolar solvent for the chemical industry, and also as a raw material for the industrial production of adipic acid and caprolactam, both of which are intermediates used in the production of nylon.

On an industrial scale, cyclohexane is produced by reacting benzene with hydrogen. Because of its unique chemical and conformational properties, cyclohexane is also used in labs in analysis and as a standard. Cyclohexane has a distinctive detergent-like odor, reminiscent of cleaning product (in which it is sometimes used).

## Structure and Physical Properties

### Conformation

The 6-vertexed ring does not conform to the shape of a perfect hexagon. The conformation of a flat 2D planar hexagon has considerable angle strain because its bonds are not 109.5 degrees; the torsional strain would also be considerable because all of the bonds would be eclipsed bonds. Therefore, to reduce torsional strain, cyclohexane adopts a three-dimensional structure known as the chair conformation. This was first proposed as early as 1890 by Hermann Sachse, but only gained widespread acceptance much later. The new conformation puts the carbons at an angle of 109.5°. Half of the hydrogens are in the plane of the ring (*equatorial*) while the other half are perpendicular to the plane (*axial*). This conformation allows for the most stable structure of cyclohexane. Another conformation of cyclohexane exists, known as boat conformation, but it interconverts to the slightly more stable chair formation. If cyclohexane is mono-substituted with a large substituent, then the substituent will most likely be found attached in an equatorial position, as this is the slightly more stable conformation.

Cyclohexane has the lowest angle and torsional strain of all the cycloalkanes, as a result cyclohexane has been deemed a 0 in total ring strain, a combination of angle and torsional strain. This also makes cyclohexane the most stable of the cycloalkanes and therefore will produce the least amount of heat (per $CH_2$ unit) when burned compared to the other cycloalkanes.

### Solid Phases

Cyclohexane has two crystalline phases. The high-temperature phase I, stable between 186 K and the melting point 280 K, is a plastic crystal, which means the molecules retain some rotational degree of freedom. The low-temperature phase II is ordered. A metastable phase III has been obtained by rapid quenching.

### Chemistry—Reactions

Pure cyclohexane in itself is rather unreactive, being a non-polar, hydrophobic hydrocarbon. It can react with very strong acids such as the superacid system HF + $SbF_5$ which will cause forced protonation and "hydrocarbon cracking". Substituted cyclohexanes, however, may be reactive under a variety of conditions, many of which are important to organic chemistry. Cyclohexane is highly flammable.

### *Derivatives*

The specific arrangement of functional groups in cyclohexane derivatives, and indeed in most cycloalkane molecules, is extremely important in chemical reactions, especially reactions involving nucleophiles. Substituents on the ring must be in the axial formation to react with other molecules. For example, the reaction of bromocyclohexane and a common nucleophile, a hydroxide anion ($OH^-$), would result in cyclohexene:

$$C_6H_{11}Br + OH^- \rightarrow C_6H_{10} + H_2O + Br^-$$

This reaction, commonly known as an elimination reaction or dehalogenation (specifically E2), requires that the bromine substituent be in the axial formation, opposing another axial H atom to react. Assuming that the bromocyclohexane was in the appropriate formation to react, the E2 reaction would commence as such:

- The electron pair bond between the C-Br moves to the Br, forming $Br^-$ and setting it free from cyclohexane
- The nucleophile (-OH) gives an electron pair to the adjacent axial H, setting H free and bonding to it to create $H_2O$
- The electron pair bond between the adjacent axial H moves to the bond between the two C-C making it C=C

Note: All three steps happen simultaneously, characteristic of all E2 reactions.

The reaction above will generate mostly E2 reactions and as a result the product will be mostly (~70%) cyclohexene. However, the percentage varies with conditions, and generally, two different reactions (E2 and $S_N2$) compete. In the above reaction, an $S_N2$ reaction would substitute the bromine for a hydroxyl ($OH^-$) group instead, but once again, the Br must be in axial to react. Once the $S_N2$ substitution is complete, the newly substituted OH group would flip back to the more stable equatorial position quickly (~1 millisecond).

### *Uses*

Commercially most of cyclohexane produced is converted into cyclohexanone–cyclohexanol mixture (or "*KA oil*") by catalytic oxidation. KA oil is then used as a raw material for adipic acid and caprolactam. Practically, if the cyclohexanol content of KA oil is higher than cyclohexanone, it is more likely (economical) to be converted into adipic acid, and the reverse case, caprolactam production is more likely. Such ratio in KA oil can be controlled by selecting suitable oxidation catalyst. Some of cyclohexane is used as an organic solvent.

Although much is already known about this cyclic hydrocarbon, research is still being done on cyclohexane and benzene mixtures and solid phase cyclohexane to determine hydrogen yields of the mix when irradiated at “195 °C.

Cyclohexane is also used for calibration of differential scanning calorimetry (DSC) instruments, because of a convenient crystal-crystal transition at “87.1 C.

Heat treating equipment manufacturer Surface Combustion uses cyclohexane as a carbon carrying gas in their high purity vacuum carburizing furnaces.

### History

Unlike compounds like benzene, cyclohexane cannot easily be obtained from natural resources such as coal. Toward the end of the nineteenth century early chemical investigators had to depend on organic synthesis. It took them 30 years to flesh out the details. In 1867 Marcellin Berthelot reduced benzene with hydroiodic acid at elevated temperatures.

He incorrectly identified the reaction product as n-hexane not only because of the convenient match in boiling point (69°C) but also because he did not believe benzene was a cyclic molecule (like his contemporary August Kekulé) but rather some sort of association of acetylene. In 1870 one of his sceptics Adolf von Baeyer repeated the reaction and pronounced the same reaction product hexahydrobenzene and in 1890 Vladimir Markovnikov believed he was able to distill the same compound from Caucasus petroleum calling his concoction hexanaphtene.

In 1894 Baeyer synthesized cyclohexane starting with a Dieckmann condensation of pimelic acid followed by multiple reductions:

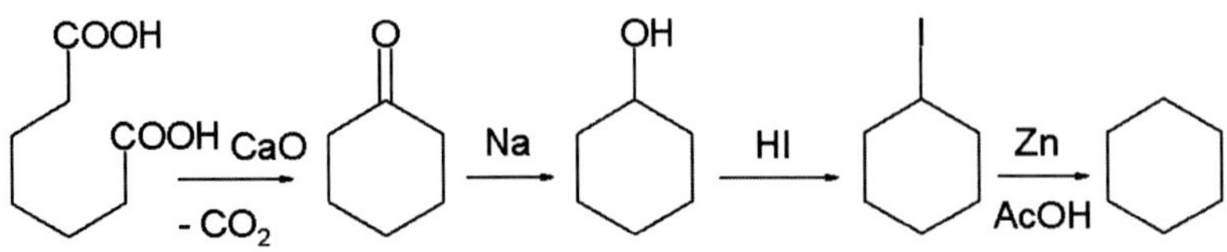


and in the same year E. Haworth and W.H. Perkin Jr. (1860–1929) did the same in a Wurtz reaction of 1,6-dibromohexane.

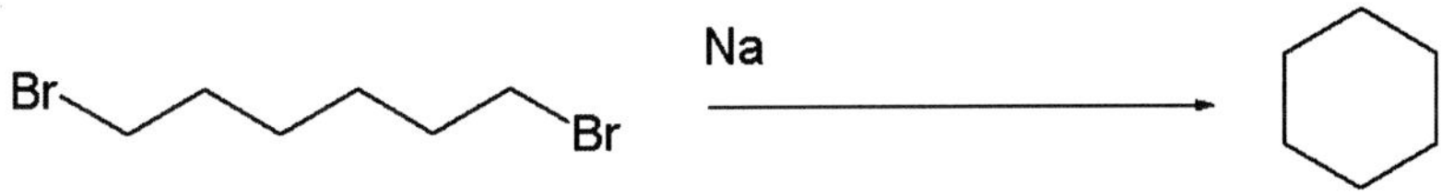


Surprisingly their cyclohexanes boiled higher by 10°C than either hexahydrobenzene or hexanaphtene but this riddle was solved in 1895 by Markovnikov, N.M.

Kishner and Nikolay Zelinsky when they re-diagnosed hexahydrobenzene and hexanaphtene as methylcyclopentane, the result of an unexpected rearrangement reaction.

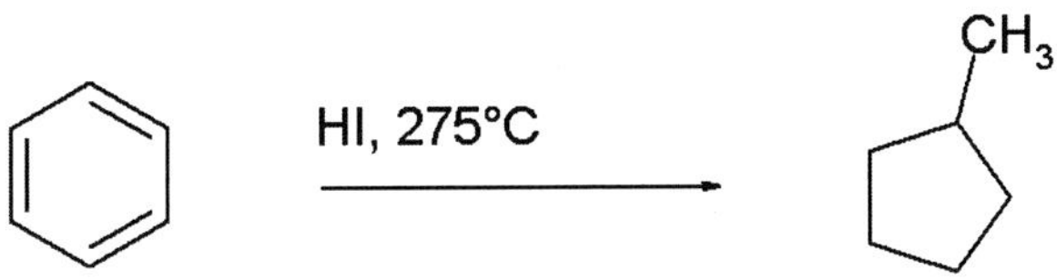

# 3

# Adipic Acid

Adipic acid is the organic compound with the formula (CH2)4(COOH)2. From the industrial perspective, it is the most important dicarboxylic acid: About 2.5 billion kilograms of this white crystalline powder are produced annually, mainly as a precursor for the production of nylon. Adipic acid otherwise rarely occurs in nature.

### *Preparation*

Historically, adipic acid was prepared from various fats using oxidation. Currently adipic acid is produced from a mixture of cyclohexanol and cyclohexanone called "KA oil", the abbreviation of "ketone-alcohol oil." The KA oil is oxidized with nitric acid to give adipic acid, via a multistep pathway. Early in the reaction the cyclohexanol is converted to the ketone, releasing nitrous acid:

$$HOC_6H_{11} + HNO_3 \rightarrow OC6H_{10} + HNO_2 + H_2O$$

Among its many reactions, the cyclohexanone is nitrosated, setting the stage for the scission of the C-C bond:

$$HNO_2 + HNO_3 \rightarrow NO + NO^{3-} + H_2O$$

$$OC_6H_{10} + NO^+ \rightarrow OC_6H^{9\text{-}2\text{-}}NO + H+$$

Side products of the method include glutaric and succinic acids.

Related processes start from cyclohexanol, which is obtained from the hydrogenation of phenol.

### *Alternative Methods of Production*

Several methods have been developed by carbonylation of butadiene. For example, the hydrocarboxylation proceeds as follows:

$$CH_2 = CHCH = CH_2 + 2\ CO + 2\ H_2O \rightarrow HO_2C(CH_2)4CO_2H$$

A method has been reported that utilizes principles of green chemistry in that water is the only by-product. Cyclohexene is oxidized with hydrogen peroxide using a tungstate-based catalyst and a phase transfer catalyst. The waste product is water.

### Uses

By far the majority of the 2.5 billion kg of adipic acid produced annually is used as monomer for the production of nylon by a polycondensation reaction with hexamethylene diamine forming 6,6-nylon. Other major applications also involve polymers: it is a monomer for production of Polyurethane and its esters are plasticizers, especially in PVC.

### In Medicine

Adipic acid has been incorporated into controlled-release formulation matrix tablets to obtain pH-independent release for both weakly basic and weakly acidic drugs. It has also been incorporated into the polymeric coating of hydrophilic monolithic systems to modulate the intragel pH, resulting in zero-order release of a hydrophilic drug. The disintegration at intestinal pH of the enteric polymer shellac has been reported to improve when adipic acid was used as a pore-forming agent without affecting release in the acidic media. Other controlled-release formulations have included adipic acid with the intention of obtaining a late-burst release profile.

### In Foods

Small but significant amounts of adipic acid are used as a food ingredient as a flavorant and gelling aid. It is used in some calcium carbonate antacids to make them tart.

### Safety

Adipic acid, like most carboxylic acids, is a mild skin irritant. It is mildly toxic, with an LD50 of 3600 mg/kg for oral ingestion by rats.

## Carboxylic Acid

Carboxylic acids are organic acids characterized by the presence of at least one carboxyl group. The general formula of a carboxylic acid is R-COOH, where R is some monovalent functional group. A carboxyl group (or carboxy) is a functional group consisting of a carbonyl (RR'C=O) and a hydroxyl (R-O-H), which has the formula -C(=O)OH, usually written as -COOH or $-CO_2H$.

Carboxylic acids are Brønsted-Lowry acids because they are proton donors. They are the most common type of organic acid. Among the simplest examples are formic acid H-COOH, that occurs in ants, and acetic acid $CH_3$-COOH, that gives vinegar its sour taste.

Acids with two or more carboxyl groups are called dicarboxylic, tricarboxylic, etc. The simplest dicarboxylic example is oxalic acid $(COOH)_2$, which is just two connected carboxyls. Mellitic acid is an example of a hexacarboxylic acid. Other important natural examples are citric acid (in lemons) and tartaric acid (in tamarinds).

Salts and esters of carboxylic acids are called carboxylates. When a carboxyl group is deprotonated, its conjugate base, a carboxylate anion is formed.

Carboxylate ions are resonance stabilized and this increased stability makes carboxylic acids more acidic than alcohols. Carboxylic acids can be seen as reduced or alkylated forms of the Lewis acid carbon dioxide; under some circumstances they can be decarboxylated to yield carbon dioxide.

## Physical Properties

### Solubility

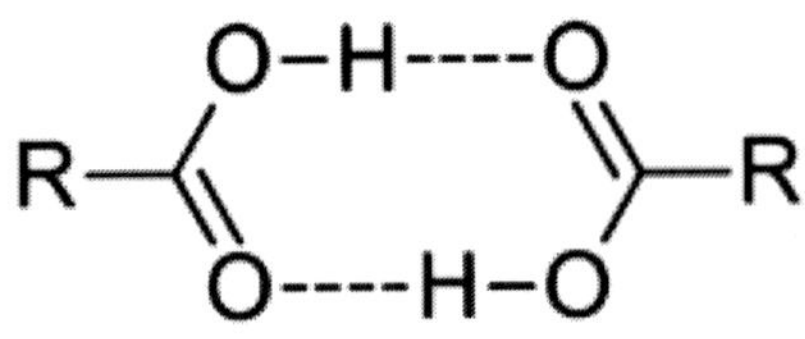


*Carboxylic Acid Dimers*

Carboxylic acids are polar. Because they are both hydrogen-bond acceptors (the carbonyl) and hydrogen-bond donors (the hydroxyl), they also participate in hydrogen bonding. Together the hydroxyl and carbonyl group forms the functional group carboxyl. Carboxylic acids usually exist as dimeric pairs in nonpolar media due to their tendency to "self-associate."

Smaller carboxylic acids (1 to 5 carbons) are soluble with water, whereas higher carboxylic acids are less soluble due to the increasing hydrophobic nature of the alkyl chain. These longer chain acids tend to be rather soluble in less-polar solvents such as ethers and alcohols.

### Boiling Points

Carboxylic acids tend to have higher boiling points than water, not only because of their increased surface area, but because of their tendency to form stabilised dimers. Carboxylic acids tend to evaporate or boil as these dimers.

For boiling to occur, either the dimer bonds must be broken, or the entire dimer arrangement must be vaporised, both of which increase enthalpy of vaporisation requirements significantly.

### Acidity

Carboxylic acids are typically weak acids, meaning that they only partially dissociate into $H^+$ cations and $RCOO^-$ anions in neutral aqueous solution.

For example, at room temperature, only 0.4% of all acetic acid molecules are dissociated. Electronegative substituents give stronger acids.

| *Carboxylic acid* | *pKa* |
|---|---|
| Formic acid ($HCO_2H$) | 3.77 |
| Acetic acid ($CH_3COOH$) | 4.76 |
| Chloroacetic acid ($CH_2ClCO_2H$) | 2.86 |
| Dichloroacetic acid ($CHCl_2CO_2H$) | 1.29 |
| Trichloroacetic acid ($CCl_3CO_2H$) | 0.65 |
| Trifluoroacetic acid ($CF_3CO_2H$) | 0.5 |
| Oxalic acid ($HO_2CCO_2H$) | 1.27 |
| Benzoic acid ($C_6H_5CO_2H$) | 4.2 |

Deprotonation of a carboxylic acid gives a carboxylate anion, which is resonance stabilized because the negative charge is delocalized between the two oxygen atoms increasing its stability. Each of the carbon-oxygen bonds in a carboxylate anion has partial double-bond character.

### Odor

Carboxylic acids often have strong odors, especially the volatile derivatives. Most common are acetic acid (vinegar) and butanoic acid (rancid butter). On the other hand, esters of carboxylic acids tend to have pleasant odors and many are used in perfumes.

### Characterization

Carboxylic acids are most readily identified as such by infrared spectroscopy. They exhibit a sharp band associated with vibration of the C-O vibration bond ($\nu_{C=O}$) between 1680 and 1725 $cm^{-1}$. A characteristic $\nu_{O\text{-}H}$ band appears as a broad peak in the 2500 to 3000 $cm^{-1}$ region. By $^1H$ NMR spectrometry, the hydroxyl hydrogen appears in the 10-13 ppm region, although it is often either broadened or not observed owing to exchange with traces of water.

### Occurrence and Applications

Many carboxylic acids are produced industrially on a large scale. They are also pervasive in nature. Esters of fatty acids are the main components of lipids and polyamides of aminocarboxylic acids are the main components of proteins.

Carboxylic acids are used in the production of polymers, pharmaceuticals, solvents, and food additives. Industrially important carboxylic acids include acetic acid (component of vinegar, precursor to solvents and coatings), acrylic and methacrylic acids (precursors to polymers, adhesives), adipic acid (polymers), citric acid (beverages), ethylenediaminetetraacetic acid (chelating agent), fatty acids (coatings), maleic acid (polymers), propionic acid (food preservative), terephthalic acid (polymers).

### Synthesis

### Industrial Routes

Industrial routes to carboxylic acids generally differ from those used on smaller scale because they require specialized equipment.

- Oxidation of aldehydes with air using cobalt and manganese catalysts. The required aldehydes are readily obtained from alkenes by hydroformylation.
- Oxidation of hydrocarbons using air. For simple alkanes, the method is nonselective but so inexpensive to be useful. Allylic and benzylic compounds undergo more selective oxidations. Alkyl groups on a benzene ring oxidized to the carboxylic acid, regardless of its chain length. Benzoic acid from toluene and terephthalic acid from para-xylene, and phthalic acid from ortho-xylene are illustrative large-scale conversions. Acrylic acid is generated from propene.
- Base-catalysed dehydrogenation of alcohols.

- Carbonylation is versatile method when coupled to the addition of water. This method is effective for alkenes that generate secondary and tertiary carbocations, e.g. isobutylene to pivalic acid. In the Koch reaction, the addition of water and carbon monoxide to alkenes is catalysed by strong acids. Acetic acid and formic acid are produced by the carbonylation of methanol, conducted with iodide and alkoxide promoters, respectively and often with high pressures of carbon monoxide, usually involving additional hydrolytic steps. Hydrocarboxylations involve the simultaneous addition of water and CO. Such reactions are sometimes called "Reppe chemistry":

$$HCCH + CO + H_2O \rightarrow CH_2{=}CHCO_2H$$

- Some long chain carboxylic acids are obtained by the hydrolysis of triglycerides obtained from plant or animal oils. These methods are related to soap making.
- fermentation of ethanol is used in the production of vinegar.

## Laboratory Methods

Preparative methods for small scale reactions for research or for production of fine chemicals often employ expensive consumable reagents.

- oxidation of primary alcohols or aldehydes with strong oxidants such as potassium dichromate, Jones reagent, potassium permanganate, or sodium chlorite. The method is amenable to laboratory conditions compared to the industrial use of air, which is "greener" since it yields less inorganic side products such as chromium or manganese oxides.
- Oxidative cleavage of olefins by ozonolysis, potassium permanganate, or potassium dichromate.
- Carboxylic acids can also be obtained by the hydrolysis of nitriles, esters, or amides, generally with acid- or base-catalysis.
- Carbonation of a Grignard and organolithium reagents:

$$RLi + CO_2\ RCO_2Li$$

$$RCO_2Li + HCl\ RCO_2H + LiCl$$

- Halogenation followed by hydrolysis of methyl ketones in the haloform reaction
- The Kolbe-Schmitt reaction provides a route to salicylic acid, precursor to aspirin.

### Less-common Reactions

Many reactions afford carboxylic acids but are used only in specific cases or are mainly of academic interest:

- Disproportionation of an aldehyde in the Cannizzaro reaction
- Rearrangement of diketones in the benzilic acid rearrangement involving the generation of benzoic acids are the von Richter reaction from nitrobenzenes and the Kolbe-Schmitt reaction from phenols.

### Reactions

The most widely practiced reactions convert carboxylic acids into esters, amides, carboxylate salts, acid chlorides, and alcohols. Carboxylic acids react with bases to form carboxylate salts, in which the hydrogen of the hydroxyl (-OH) group is replaced with a metal cation. Thus, acetic acid found in vinegar reacts with sodium bicarbonate (baking soda) to form sodium acetate, carbon dioxide, and water:

$$CH_3COOH + NaHCO_3 \rightarrow CH_3COO^-Na^+ + CO_2 + H_2O$$

Carboxylic acids also react with alcohols to give esters. This process is heavily used in the production of polyesters. Similarly carboxylic acids are converted into amides, but this conversion typically does not occur by direct reaction of the carboxylic acid and the amine. Instead esters are typical precursors to amides. The conversion of amino acids into peptides is a major biochemical process that requires ATP. The hydroxyl group on carboxylic acids may be replaced with a chlorine atom using thionyl chloride to give acyl chlorides. In nature, carboxylic acids are converted to thioesters.

Carboxylic acid can be reduced to the alcohol by hydrogenation or using stoichiometric hydride reducing agents such as lithium aluminium hydride. N,N-dimethylchloromethylenammonium chloride is a highly chemoselective agent for carboxylic acid reduction. It selectively activate the carboxylic acid and is known to tolerate active functionalities such as ketone as well as the moderate ester, olefin, nitrile and halide moeties.

### Specialized Reactions

- As with all carbonyl compounds, the protons on the α-carbon are labile due to keto-enol tautomerization. Thus the α-carbon is easily halogenated in the Hell-Volhard-Zelinsky halogenation.

- The Schmidt reaction converts carboxylic acids to amines.
- Carboxylic acids are decarboxylated in the Hunsdiecker reaction.
- The Dakin-West reaction converts an amino acid to the corresponding amino ketone.
- In the Barbier-Wieland degradation, the alpha-methylene group in an aliphatic carboxylic acid is removed in a sequence of reaction steps, effectively a chain-shortening. The inverse procedure is the Arndt-Eistert synthesis, where an acid is converted into acyl halide and reacts with diazomethane to give the highest homolog.
- Many acids undergo decarboxylation. Enzymes that catalyze these reactions are known as carboxylases (EC 6.4.1) and decarboxylases (EC 4.1.1).
- Carboxylic acids are reduced to aldehydes via the ester and DIBAL, via the acid chloride in the Rosenmund reduction and via the thioester in the Fukuyama reduction.

### Nomenclature and Examples

Although rarely used, IUPAC-recommended names also exist. For example, butyric acid ($C_3H_7CO_2H$) is, according to IUPAC guidelines, also known as butanoic acid.

The carboxylate anion R-COO$^-$ is usually named with the suffix *-ate*, so acetic acid, for example, becomes acetate ion. In IUPAC nomenclature, carboxylic acids have an *-oic acid* suffix (e.g., octadecanoic acid). In common nomenclature, the suffix is usually -*ic acid* (e.g., stearic acid).

*Table 1 : Straight-chained, saturated carboxylic acids*

| **Carbon atoms** | **Common name** | **IUPAC name** | **Chemical formula** | **Common location or use** |
|---|---|---|---|---|
| 1 | Formic acid | Methanoic acid | HCOOH | Insect stings |
| 2 | Acetic acid | Ethanoic acid | $CH_3COOH$ | Vinegar |
| 3 | Propionic acid | Propanoic acid | $CH_3CH_2COOH$ | Preservative for stored grains |
| 4 | Butyric acid | Butanoic acid | $CH_3(CH_2)_2COOH$ | Rancid butter |
| 5 | Valeric acid | Pentanoic acid | $CH_3(CH_2)_3COOH$ | Valerian |
| 6 | Caproic acid | Hexanoic acid | $CH_3(CH_2)_4COOH$ | Goat fat |
| 7 | Enanthic acid | Heptanoic acid | $CH_3(CH_2)_5COOH$ | |

| | | | | |
|---|---|---|---|---|
| 8 | Caprylic acid | Octanoic acid | $CH_3(CH_2)_6COOH$ | Coconuts and breast milk |
| 9 | Pelargonic acid | Nonanoic acid | $CH_3(CH_2)_7COOH$ | Pelargonium |
| 10 | Capric acid | Decanoic acid | $CH_3(CH_2)_8COOH$ | |
| 12 | Lauric acid | Dodecanoic acid | $CH_3(CH_2)_{10}COOH$ | Coconut oil and hand wash soaps. |
| 14 | Myristic acid | Tetradecanoic acid | $CH_3(CH_2)_{12}COOH$ | Nutmeg |
| 16 | Palmitic acid | Hexadecanoic acid | $CH_3(CH_2)_{14}COOH$ | Palm oil |
| 18 | Stearic acid | Octadecanoic acid | $CH_3(CH_2)_{16}COOH$ | Chocolate, waxes, soaps, and oils |
| 20 | Arachidic acid | Icosanoic acid | $CH_3(CH_2)_{18}COOH$ | Peanut oil |

*Table 2 : Other Carboxylic Acids*

| ***Compound class*** | ***Members*** |
|---|---|
| unsaturated monocarboxylic acids | acrylic acid (2-propenoic acid) – $CH_2$=CHCOOH, used in polymer synthesis |
| Fatty acids | medium to long-chain saturated and unsaturated monocarboxylic acids, with even number of carbons examples docosahexaenoic acid and eicosapentaenoic acid (nutritional supplements) |
| Amino acids | the building blocks of proteins |
| Keto acids | acids of biochemical significance that contain a ketone group e.g. acetoacetic acid and pyruvic acid |
| Aromatic carboxylic | benzoic acid, the sodium salt of benzoic acid is used as |
| acids | a food preservative, salicylic acid – a beta hydroxy type found in many skin care products |
| Dicarboxylic acids | containing two carboxyl groups examples adipic acid the monomer used to produce nylon and aldaric acid – a family of sugar acids |
| Tricarboxylic acids | containing three carboxyl groups example citric acid – found in citrus fruits and isocitric acid |
| Alpha hydroxy acids | containing a hydroxy group example glyceric acid, glycolic acid and lactic acid (2-hydroxypropanoic acid) – found in sour milk tartaric acid - found in wine. |

## Carboxyl Radical

The radical ·COOH (CAS# 2564-86-5) has only a separate fleeting existence. The acid dissociation constant of ·COOH has been measured using electron paramagnetic resonance spectrocopy. The carboxyl group tends to dimerise to form oxalic acid.

## Nylon

Nylon is a generic designation for a family of synthetic polymers known generically as polyamides, first produced on February 28, 1935, by Wallace Carothers at DuPont's research facility at the DuPont Experimental Station. Nylon is one of the most commonly used polymers.

### *Overview*

Nylon is a thermoplastic, silky material, first used commercially in a nylon-bristled toothbrush (1938), followed more famously by women's stockings ("nylons"; 1940). It is made of repeating units linked by amide bonds and is frequently referred to as *polyamide* (PA). Nylon was the first commercially successful synthetic polymer.

There are two common methods of making nylon for fibre applications. In one approach, molecules with an acid (-COOH) group on each end are reacted with molecules containing amine (-$NH_2$) groups on each end. The resulting nylon is named on the basis of the number of carbon atoms separating the two acid groups and the two amines. These are formed into monomers of intermediate molecular weight, which are then reacted to form long polymer chains.

Nylon was intended to be a synthetic replacement for silk and substituted for it in many different products after silk became scarce during World War II. It replaced silk in military applications such as parachutes and flak vests, and was used in many types of vehicle tires.

Nylon fibres are used in many applications, including fabrics, bridal veils, carpets, musical strings, and rope.

Solid nylon is used for mechanical parts such as machine screws, gears and other low- to medium-stress components previously cast in metal. Engineering-grade nylon is processed by extrusion, casting, and injection molding. Solid nylon is used in hair combs. Type 6,6 Nylon 101 is the most common commercial grade of nylon, and Nylon 6 is the most common commercial grade of molded nylon. For use in tools such as the Apple Black Stick, a nylon is available in glass-filled variants which increase structural and impact strength and rigidity, and molybdenum sulfide-filled variants which increase lubricity.

Aramids are another type of polyamide with quite different chain structures which include aromatic groups in the main chain. Such polymers make excellent ballistic fibres.

## Chemistry

Nylons are condensation copolymers formed by reacting equal parts of a diamine and a dicarboxylic acid, so that amides are formed at both ends of each monomer in a process analogous to polypeptide biopolymers. Chemical elements included are carbon, hydrogen, nitrogen, and oxygen. The numerical suffix specifies the numbers of carbons donated by the monomers; the diamine first and the diacid second.

The most common variant is nylon 6-6 which refers to the fact that the diamine (hexamethylene diamine, IUPAC name: hexane-1,6-diamine) and the diacid (adipic acid, IUPAC name: hexanedioic acid) each donate 6 carbons to the polymer chain. As with other regular copolymers like polyesters and polyurethanes, the "repeating unit" consists of one of each monomer, so that they alternate in the chain. Since each monomer in this copolymer has the same reactive group on both ends, the direction of the amide bond reverses between each monomer, unlike natural polyamide proteins which have overall directionality: C terminal → N terminal. In the laboratory, nylon 6-6 can also be made using adipoyl chloride instead of adipic.

It is difficult to get the proportions exactly correct, and deviations can lead to chain termination at molecular weights less than a desirable 10,000 daltons (u). To overcome this problem, a crystalline, solid "nylon salt" can be formed at room temperature, using an exact 1:1 ratio of the acid and the base to neutralize each other. Heated to 285 °C (545 °F), the salt reacts to form nylon polymer. Above 20,000 daltons, it is impossible to spin the chains into yarn, so to combat this, some acetic acid is added to react with a free amine end group during polymer elongation to limit the molecular weight. In practice, and especially for 6,6, the monomers are often combined in a water solution. The water used to make the solution is evaporated under controlled conditions, and the increasing concentration of "salt" is polymerized to the final molecular weight.

DuPont patented nylon 6,6, so in order to compete, other companies (particularly the German BASF) developed the homopolymer nylon 6, or polycaprolactam — not a condensation polymer, but formed by a ring-opening polymerization (alternatively made by polymerizing aminocaproic acid). The peptide bond within the caprolactam is broken with the exposed active groups on each side being incorporated into two new bonds as the monomer becomes part of the polymer backbone.

In this case, all amide bonds lie in the same direction, but the properties of nylon 6 are sometimes indistinguishable from those of nylon 6,6 — except for melt temperature and some fibre properties in products like carpets and textiles. There is also nylon 9.

The 428 °F (220 °C) melting point of nylon 6 is lower than the 509 °F (265 °C) melting point of nylon 6,6.

Nylon 5,10, made from pentamethylene diamine and sebacic acid, was studied by Carothers even before nylon 6,6 and has superior properties, but is more expensive to make. In keeping with this naming convention, "nylon 6,12" (N-6,12) or "PA-6,12" is a copolymer of a 6C diamine and a 12C diacid. Similarly for N-5,10 N-6,11; N-10,12, etc.

Other nylons include copolymerized dicarboxylic acid/diamine products that are *not* based upon the monomers listed above. For example, some aromatic nylons are polymerized with the addition of diacids like terephthalic acid (→ Kevlar Twaron) or isophthalic acid (→ Nomex), more commonly associated with polyesters. There are copolymers of N-6,6/N6; copolymers of N-6,6/N-6/N-12; and others. Because of the way polyamides are formed, nylon would seem to be limited to unbranched, straight chains. But "star" branched nylon can be produced by the condensation of dicarboxylic acids with polyamines having three or more amino groups.

The general reaction is:

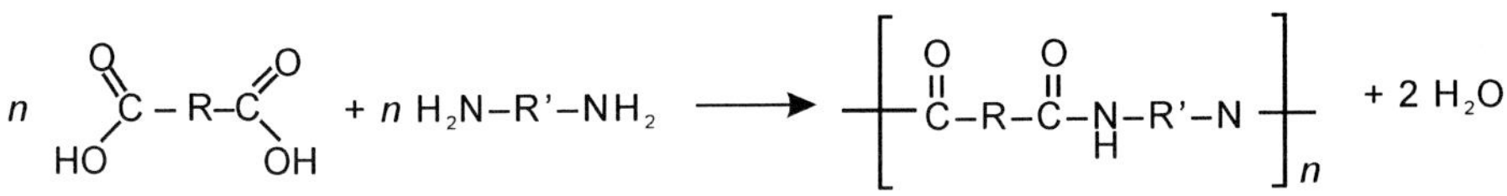

A molecule of water is given off and the nylon is formed. Its properties are determined by the R and R' groups in the monomers. In nylon 6,6, R = 4C and R' = 6C alkanes, but one also has to include the two carboxyl carbons in the diacid to get the number it donates to the chain. In Kevlar, both R and R' are benzene rings.

## Concepts of Nylon Production

The first approach: combining molecules with an acid (COOH) group on each end are reacted with two chemicals that contain amine ($NH_2$) groups on each end. This process creates nylon 6,6, made of hexamethylene diamine with six carbon atoms and adipic acid. The second approach: a compound has an acid at one end and an amine at the other and is polymerized to form a chain with repeating units of (-NH-$[CH_2]_n$-CO-$)_x$. In other words, nylon 6 is made from a single six-carbon substance called caprolactam. In this equation, if n = 5, then nylon 6 is the assigned name (may also be referred to as polymer).

The characteristic features of nylon 6,6 include:

- Pleats and creases can be heat-set at higher temperatures
- More compact molecular structure
- Better weathering properties; better sunlight resistance
- Softer "Hand"
- Higher melting point (256 °C/492.8 °F)
- Superior colorfastness
- Excellent abrasion resistance.

On the other hand, nylon 6 is easy to dye, more readily fades; it has a higher impact resistance, a more rapid moisture absorption, greater elasticity and elastic recovery.

### *Characteristics*

- Variation of luster: nylon has the ability to be very lustrous, semilustrous or dull.
- Durability: its high tenacity fibres are used for seatbelts, tire cords, ballistic cloth and other uses.
- High elongation
- Excellent abrasion resistance
- Highly resilient (nylon fabrics are heat-set)
- Paved the way for easy-care garments
- High resistance to insects, fungi, animals, as well as molds, mildew, rot and many chemicals
- Used in carpets and nylon stockings
- Melts instead of burning
- Used in many military applications
- Good specific strength
- Transparent to infrared light ("12dB).

### *Bulk Properties*

Above their melting temperatures, $T_m$, thermoplastics like nylon are amorphous solids or viscous fluids in which the chains approximate random coils. Below $T_m$, amorphous regions alternate with regions which are lamellar crystals. The amorphous regions contribute elasticity and the crystalline regions contribute strength and rigidity. The planar amide (-CO-NH-) groups are very polar, so nylon forms multiple hydrogen bonds among adjacent strands. Because the nylon backbone is so regular and symmetrical, especially if all the amide bonds are in the *trans* configuration, nylons often have high crystallinity and make excellent fibres. The amount of crystallinity depends on the details of formation, as well as on the kind of nylon. Apparently it can never be quenched from a melt as a completely amorphous solid.

Nylon 6,6 can have multiple parallel strands aligned with their neighboring peptide bonds at coordinated separations of exactly 6 and 4 carbons for considerable lengths, so the carbonyl oxygens and amide hydrogens can line up to form interchain hydrogen bonds repeatedly, without interruption. Nylon 5,10 can have coordinated runs of 5 and 8 carbons. Thus parallel (but not antiparallel) strands can participate in extended, unbroken, multi-chain β-pleated sheets, a strong and tough supermolecular structure similar to that found in natural silk fibroin and the β-keratins in feathers. (Proteins have only an amino acid α-carbon separating sequential -CO-NH- groups.) Nylon 6 will form uninterrupted H-bonded sheets with mixed directionalities, but the β-sheet wrinkling is somewhat different. The three-dimensional disposition of each alkane hydrocarbon chain depends on rotations about the 109.47° tetrahedral bonds of singly bonded carbon atoms.

When extruded into fibres through pores in an industrial spinneret, the individual polymer chains tend to align because of viscous flow. If subjected to cold drawing afterwards, the fibres align further, increasing their crystallinity, and the material acquires additional tensile strength. In practice, nylon fibres are most often drawn using heated rolls at high speeds.

Block nylon tends to be less crystalline, except near the surfaces due to shearing stresses during formation. Nylon is clear and colorless, or milky, but is easily dyed. Multistranded nylon cord and rope is slippery and tends to unravel. The ends can be melted and fused with a heat source such as a flame or electrode to prevent this.

When dry, polyamide is a good electrical insulator. However, polyamide is hygroscopic. The absorption of water will change some of the material's properties such as its electrical resistance. Nylon is less absorbent than wool or cotton.

### *Historical Uses*

Bill Pittendreigh, DuPont, and other individuals and corporations worked diligently during the first few months of World War II to find a way to replace Asian silk and hemp with nylon in parachutes. It was also used to make tires, tents, ropes, ponchos, and other military supplies. It was even used in the production of a high-grade paper for U.S. currency. At the outset of the war, cotton accounted for more than 80% of all fibres used and manufactured, and wool fibres accounted for nearly all of the rest. By August 1945, manufactured fibres had taken a market share of 25%, at the expense of cotton.

Some of the terpolymers based upon nylon are used every day in packaging. Nylon has been used for meat wrappings and sausage sheaths.

### *Use in Composites*

Nylon can be used as the matrix material in composite materials, with reinforcing fibres like glass or carbon fibre; such a composite has a higher density than pure nylon. Such thermoplastic composites (25% glass fibre) are frequently used in car components next to the engine, such as intake manifolds, where the good heat resistance of such materials makes them feasible competitors to metals.

### *Hydrolysis and Degradation*

All nylons are susceptible to hydrolysis, especially by strong acids, a reaction essentially the reverse of the synthetic reaction shown above. The molecular weight of nylon products so attacked drops fast, and cracks form quickly at the affected zones. Lower members of the nylons (such as nylon 6) are affected more than higher members such as nylon 12. This means that nylon parts cannot be used in contact with sulfuric acid for example, such as the electrolyte used in lead-acid batteries. When being molded, nylon must be dried to prevent hydrolysis in the molding machine barrel since water at high temperatures can also degrade the polymer. The reaction is of the type:

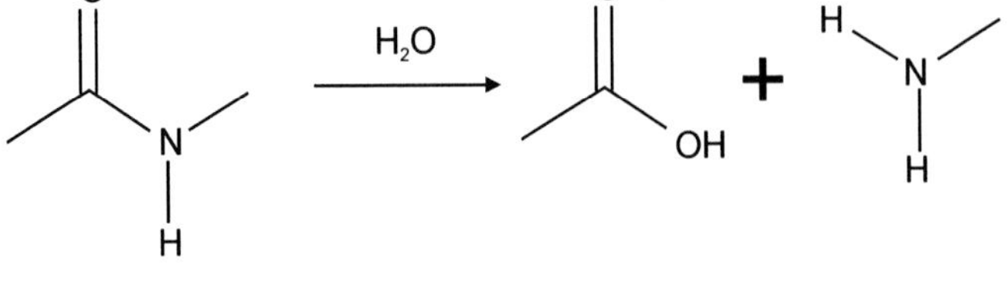

### *Incineration and Recycling*

Various nylons break down in fire and form hazardous smoke, and toxic fumes or ash, typically containing hydrogen cyanide.

Incinerating nylons to recover the high energy used to create them is usually expensive, so most nylons reach the garbage dumps, decaying very slowly. Some recycling is done on nylon, usually creating pellets for reuse in the industry.

### *Etymology*

In 1940, John W. Eckelberry of DuPont stated that the letters "nyl" were arbitrary and the "on" was copied from the suffixes of other fibres such as cotton and rayon.

A later publication by DuPont explained that the name was originally intended to be "No-Run" ("run" meaning "unravel"), but was modified to avoid making such an unjustified claim and to make the word sound better.

An apocryphal explanation is that Nylon is a conflation of "New York" and "London". A humorous backronym is "Now You've Lost, Old Nippon" referring to the supposed loss of demand for Japanese silk.

## Polyamide

A polyamide is a polymer containing monomers of amides joined by peptide bonds.

They can occur both naturally and artificially, examples being proteins, such as wool and silk, and can be made artificially through step-growth polymerization or solid-phase synthesis, examples being nylons, aramids, and sodium poly(aspartate). Polyamides are commonly used in textiles, automotives, carpet and sportswear due to their extreme durability and strength.

### *Classification*

According to the number of repeating units' types, polyamides can be:

- homopolymers :
  - o PA 6 : $[NH–(CH_2)_5–CO]_n$ made from ε-Caprolactam ;
  - o PA 66 : $[NH–(CH_2)_6–NH–CO–(CH_2)_4–CO]_n$ made from hexamethylenediamine and adipic acid;
- copolymers :

- o PA 6/66 : [NH-$(CH_2)_6$–NH–CO–$(CH_2)_4$–CO]$_n$–[NH–$(CH_2)_5$–CO]$_m$ made from caprolactam, hexamethylenediamine and adipic acid ;
- o PA 66/610 : [NH–$(CH_2)_6$–NH–CO–$(CH_2)_4$–CO]$_n$–[NH–$(CH_2)_6$–NH–CO–$(CH_2)_8$–CO]$_m$ made from hexamethylenediamine, adipic acid and sebacic acid.

According to their crystallinity, polyamides can be:

- semi-crystalline:
  - o high crystallinity : PA46 et PA 66 ;
  - o low crystallinity : PA mXD6 made from m-xylylenediamine and adipic acid;
- amorphous : PA 6I made from hexamethylenediamine and isophthalic acid.

According to this classification, PA66, for example, is an aliphatic semi-crystalline homopolyamide.

### *Production from Monomers*

The amide link is produced from the condensation reaction of an amino group and a carboxylic acid or acid chloride group. A small molecule, usually water, or hydrogen chloride, is eliminated.

The amino group and the carboxylic acid group can be on the same monomer, or the polymer can be constituted of two different bifunctional monomers, one with two amino groups, the other with two carboxylic acid or acid chloride groups.

Amino acids can be taken as examples of single monomer (if the difference between R groups is ignored) reacting with identical molecules to form a polyamide:

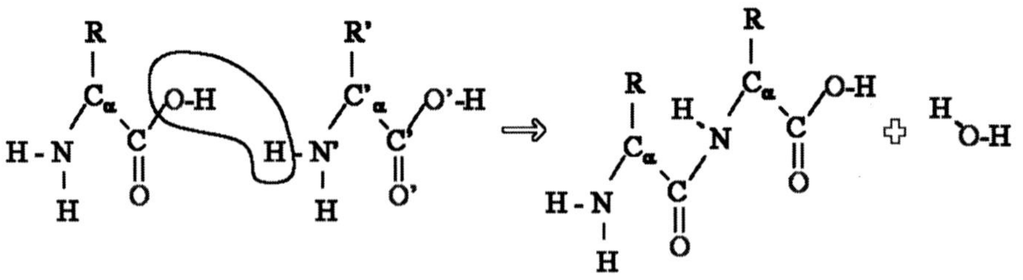

*The reaction of two amino acids. Many of these reactions produce long chain proteins*

Aramid is made from two different monomers which continuously alternate to form the polymer and is an aromatic polyamide:

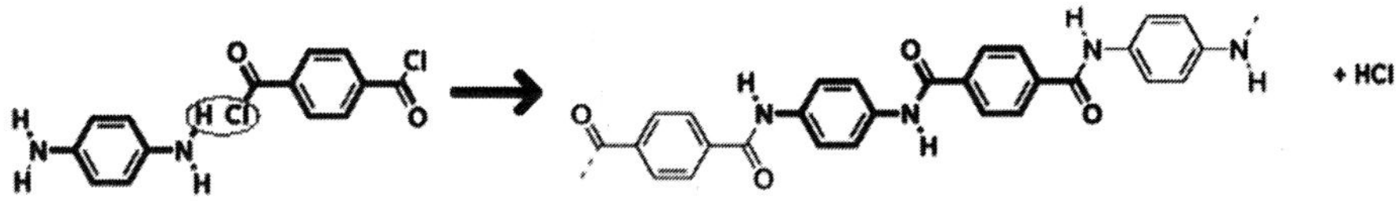


*The reaction of 1,4-phenyl-diamine (para-phenylenediamine) and terephthaloyl chloride to produce Aramid*

## *Antistatic Need*

These materials are commonly completely insulating, and generate static electricity, which can cause dangerous sparks that can damage electronics in manufacturing plants and in consumer products. This has resulted in an interest in incorporation of conductive fillers such as carbon black, metals and conducting polymers. The most common conducting fillers are silver and carbon black. Both of these materials have processing deficiencies, while silver is also prohibitively expensive for applications such as antistatics.

## Caprolactam

Caprolactam is an organic compound with the formula $(CH_2)_5C(O)NH$. This colourless solid is a lactam or a cyclic amide of caproic acid. Approximately 2 billion kilograms are produced annually. Caprolactam is the precursor to Nylon 6, a widely used synthetic polymer.

## *Synthesis and Production*

Caprolactam was first described in the late 1800s when it was prepared by the cyclization of ε-aminocaproic acid, the product of the hydrolysis of caprolactam. Given the commercial significance of nylon-6, many methods have been developed for the production of caprolactam. Most of the caprolactam is synthesised from cyclohexanone (1), which is first converted to its oxime (2). Treatment of this oxime with acid induces the Beckmann rearrangement to give caprolactam (3):

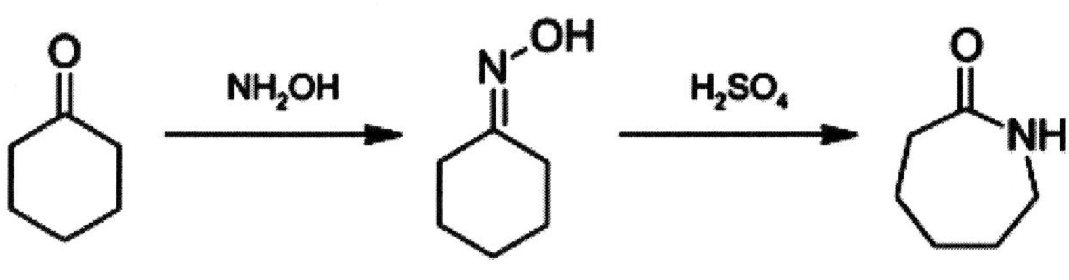


**1** **2** **3**

The immediate product of the acid-induced rearrangement is the bisulfate salt of caprolactam. This salt is neutralized with ammonia to release the free lactam and cogenerate ammonium sulfate. In optimizing the industrial practices, much attention is directed toward minimizing the production of ammonium salts.

The other major industrial route involves formation of the oxime from cyclohexane using nitrosyl chloride. The advantage of this method is that cyclohexane is less expensive than cyclohexanone. In earlier times, caprolactam was prepared by treatment of caprolactone with ammonia.

### Uses

Caprolactam is the precursor to Nylon-6. The conversion entails a ring-opening polymerization:

$$n\ (CH_2)_5C(O)NH \rightarrow [(CH_2)_5C(O)NH]_n$$

Nylon-6 is widely used in fibres and plastics.

### Safety

Caprolactam is an irritant and is mildly toxic, with an $LD_{50}$ of 1.1 g/kg (rat, oral). In 1991, it was included on the list of hazardous air pollutants by the U.S. Clean Air Act of 1990. It was subsequently removed from the list in 1993. In water, caprolactam hydrolyzes to aminocaproic acid, which is used medicinally.

As of 2011 caprolactam had the unusual status of being the only chemical in the International Agency for Research on Cancer's lowest hazard category, Group 4 "probably not carcinogenic to humans".

## Amide

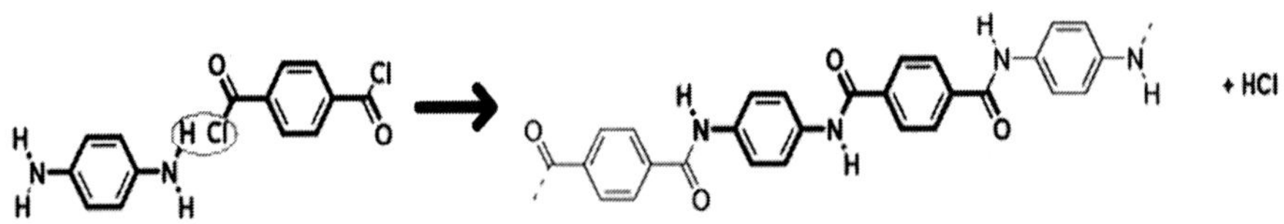


*Amide functional group*

In chemistry, an amide is an organic compound that contains the functional group consisting of a carbonyl group (R-C=O) linked to a nitrogen atom (N). The term refers both to a class of compounds and a functional group within those compounds. The term *amide* also refers to deprotonated form of ammonia ($NH_3$) or an amine, often represented as anions $R_2N^-$. The remainder of this article is about the carbonyl-nitrogen sense of *amide*.

## Structure and Bonding

The simplest amides are derivatives of ammonia wherein one hydrogen atom has been replaced by an acyl group. The ensemble is generally represented as $RC(O)NH_2$. Closely related and even more numerous are amides derived from primary amines ($R'NH_2$) with the formula RC(O)NHR'.

Amides are also commonly derived from secondary amines (R'R*NH) with the formula RC(O)NR'R*. Amide are usually regarded as derivatives of carboxylic acids in which the hydroxyl group has been replaced by an amine or ammonia.

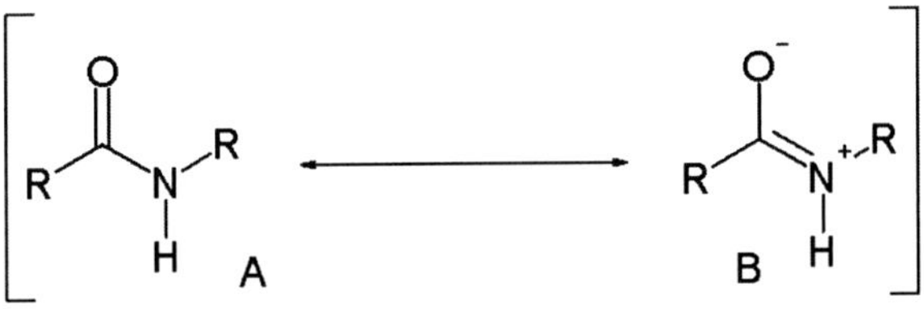


The lone pair of electrons on the nitrogen is delocalized into the carbonyl, thus forming a partial double bond between N and the carbonyl carbon. Consequently the nitrogen in amides is not pyramidal. It is estimated that acetamide is described by resonance structure A for 62% and by B for 28%

## Nomenclature

In the usual nomenclature, one adds the term "amide" to the stem of the parent acid's name. For instance, the amide derived from acetic acid is named acetamide ($CH_3CONH_2$). IUPAC recommends ethanamide, but this and related formal names are rarely encountered. When the amide is derived from a primary or secondary amine, the substitutents on nitrogen are indicated first in the name.

Thus, the amide formed from dimethylamine and acetic acid is *N,N*-dimethylacetamide ($CH_3CONMe_2$, where Me = $CH_3$). Usually even this name is simplified to dimethylacetamide. Cyclic amides are called lactams; they are necessarily secondary or tertiary amides. Functional groups consisting of $-P(O)NR_2$ and $-SO_2NR_2$ are phosphonamides and sulfonamides, respectively.

### Properties—Basicity

Compared to amines, amides are very weak bases. While the conjugate acid of an amine has a pKa of about 9.5, the conjugate acid of an amide has a pKa around -0.5. Therefore amides don't have as clearly noticeable acid-base properties in water. This lack of basicity is explained by the electron-withdrawing nature of the carbonyl group where the lone pair of electrons on the nitrogen is delocalized by resonance.

On the other hand, amides are much stronger bases than carboxylic acids, esters, aldehydes, and ketones (conjugated acid pKa between -6 and -10). It is estimated in silico that acetamide is represented by resonance structure A for 62% and by B for 28%. Resonance is largely prevented in the very strained quinuclidone.

Because of the greater electronegativity of oxygen, the carbonyl (C=O) is a stronger dipole than the N-C dipole. The presence of a C=O dipole and, to a lesser extent a N-C dipole, allows amides to act as H-bond acceptors. In primary and secondary amides, the presence of N-H dipoles allows amides to function as H-bond donors as well. Thus amides can participate in hydrogen bonding with water and other protic solvents; the oxygen atom can accept hydrogen bonds from water and the N-H hydrogen atoms can donate H-bonds. As a result of interactions such as these, the water solubility of amides is greater than that of corresponding hydrocarbons.

The proton of a primary or secondary amide does not dissociate readily under normal conditions; its $pK_a$ is usually well above 15. Conversely, under extremely acidic conditions, the carbonyl oxygen can become protonated with a $pK_a$ of roughly –1.

### Solubility

The solubilities of amides and esters are roughly comparable. Typically amides are less soluble than comparable amines and carboxylic acids since these compounds can both donate and accept hydrogen bonds. Tertiary amides, with the important exception of N,N-dimethylformamide, exhibit low solubility in water.

### Characterization

The presence of the functional group is generally easily established, at least in small molecules. They are the most common non-basic functional group. They can be distinguished from nitro and cyano groups by their IR spectra. Amides exhibit a moderately intense $\nu_{CO}$ band near 1650 $cm^{-1}$. By $^1H$ NMR spectroscopy, CON*H*R signals occur at low fields. In X-ray crystallography, the C(O)N centre together with the three immediately adjacent atoms characteristically define a plane.

### Applications and Occurrence

Amides are pervasive in nature and technology as structural materials. The amide linkage is easily formed, confers structural rigidity, and resists hydrolysis.

Nylons are polyamides as are the very resilient materials Aramid, Twaron, and Kevlar. Amide linkages in a biochemical context are called peptide linkages. Amide linkages constitute a defining molecular feature of proteins, the secondary structure of which is due in part to the hydrogen bonding abilities of amides. Low molecular weight amides, such as dimethylformamide ($HC(O)N(CH_3)_2$), are common solvents. Many drugs are amides, including penicillin and LSD.

### Amide Synthesis

Amides are commonly formed via reactions of a carboxylic acid with an amine. Many methods are known for driving the unfavorable equilibrium to the right:

$$RCO_2H + R'R''NH \rightleftharpoons RC(O)NR'R^- + H_2O$$

For the most part, these reactions involve "activating" the carboxylic acid and the best known method, the Schotten-Baumann reaction, which involves conversion of the acid to the acid chlorides:

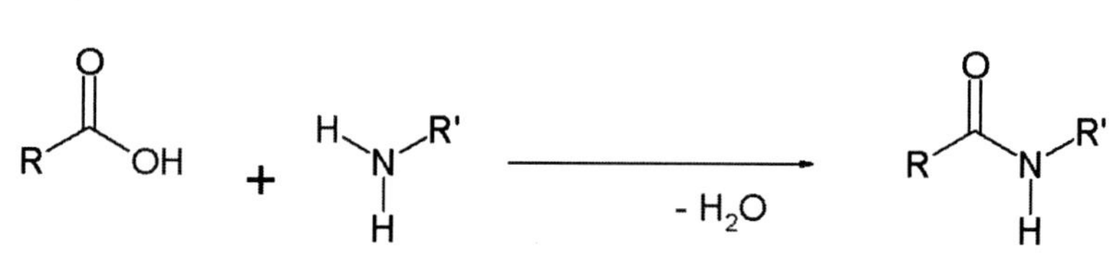


### Amide Reactions

Amides undergo many chemical reactions, usually through an attack on the carbonyl breaking the carbonyl double bond and forming a tetrahedral intermediate. Thiols, hydroxyls and amines are all known to serve as nucleophiles. Owing to their resonance stabilization, amides are less reactive under physiological conditions than esters. Enzymes, e.g. peptidases or artificial catalysts, are known to accelerate the hydrolysis reactions. They can be hydrolysed in hot alkali, as well as in strong acidic conditions. Acidic conditions yield the carboxylic acid and the ammonium ion while basic hydrolysis yield the carboxylate ion and ammonia. Amides are also versatile precursors to many other functional groups.

| ***Reaction name*** | ***Product*** | ***Comment*** |
|---|---|---|
| dehydration | nitrile | reagent: phosphorus pentoxide |
| Hofmann rearrangement | amine with one fewer carbon atoms | reagents: bromine and sodium hydroxide |
| amide reduction | amine | reagent: lithium aluminium hydride |
| Vilsmeier-Haack reaction | imine | $POCl_3$, aromatic substrate, formamide |

# 4

# Nitrobenzene

Nitrobenzene is an organic compound with the chemical formula $C_6H_5NO_2$. It is a water-insoluble pale yellow oil with an almond-like odor. It freezes to give greenish-yellow crystals. It is produced on a large scale as a precursor to aniline. Although occasionally used as a flavouring or perfume additive, nitrobenzene is highly toxic in large quantities. In the laboratory, it is occasionally used as a solvent, especially for electrophilic reagents.

### *Production*

Nitrobenzene is prepared by nitration of benzene with a mixture of concentrated sulfuric acid, water, and nitric acid. This mixture is sometimes called "mixed acid."

The production of nitrobenzene is one of the more dangerous processes conducted in the chemical industry because of the exothermicity of the reaction ($\Delta H$ = –117 kJ/mol).

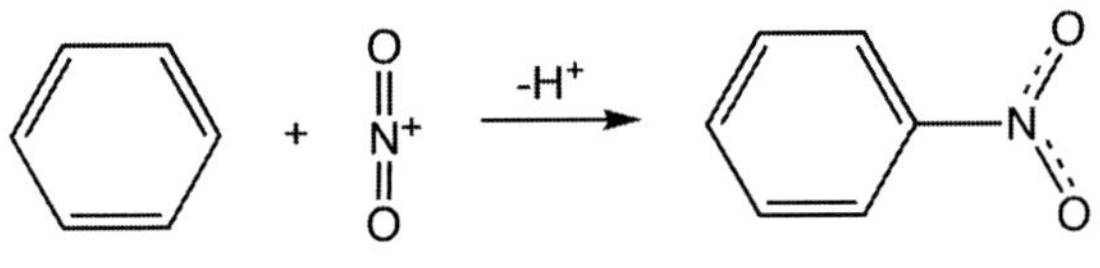


World capacity for nitrobenzene in 1985 was about $1.7 \times 10^6$ tonnes.

### *Mechanism of Nitration*

The reaction pathway entails formation of an adduct between the Lewis acidic nitronium ion ($NO_2^+$) and benzene. The nitronium ion is generated in situ by the reaction of nitric acid and an acidic dehydration agent, typically sulfuric acid:

$$HNO_3 + H^+ \rightleftharpoons NO_2^+ + H_2O$$

### Uses

Approximately 95% of nitrobenzene is consumed in the production of aniline, which is a precursor to rubber chemicals, pesticides, dyes, explosives, and pharmaceuticals.

### Specialized Applications

Nitrobenzene is also used in shoe and floor polishes, leather dressings, paint solvents, and other materials to mask unpleasant odors. Redistilled, as oil of mirbane, nitrobenzene has been used as an inexpensive perfume for soaps. A significant merchant market for nitrobenzene is its use in the production of the analgesic paracetamol (also known as acetaminophen) (Mannsville 1991). Nitrobenzene is also used in Kerr cells, as it has an unusually large Kerr constant.

### Organic Reactions

Aside from its conversion to aniline, nitrobenzene is readily converted to related derivatives such as azobenzene, nitrosobenzene, and phenylhydroxylamine. The nitro- group is deactivating, thus substitution tends to occur at the *meta*-position.

### Safety

Nitrobenzene is highly toxic (Threshold Limit Value 5 $mg/m^3$) and readily absorbed through the skin.

Prolonged exposure may cause serious damage to the central nervous system, impair vision, cause liver or kidney damage, anemia and lung irritation. Inhalation of fumes may induce headache, nausea, fatigue, dizziness, cyanosis, weakness in the arms and legs, and in rare cases may be fatal. The oil is readily absorbed through the skin and may increase heart rate, cause convulsions or rarely death. Ingestion may similarly cause headaches, dizziness, nausea, vomiting and gastrointestinal irritation, loss of limbs and also causes internal bleeding.

Nitrobenzene is considered a likely human carcinogen by the United States Environmental Protection Agency, and is classified by the IARC as a Group 2B carcinogen which is "possibly carcinogenic to humans". It has been shown to cause liver, kidney, and thyroid adenomas and carcinomas in rats.

## Aniline

Aniline, phenylamine or aminobenzene is an organic compound

with the formula $C_6H_5NH_2$. Consisting of a phenyl group attached to an amino group, aniline is the prototypical aromatic amine. Being a precursor to many industrial chemicals, its main use is in the manufacture of precursors to polyurethane. Like most volatile amines, it possesses the somewhat unpleasant odour of rotten fish. It ignites readily, burning with a smoky flame characteristic of aromatic compounds. Aniline is colorless, but it slowly oxidizes and resinifies in air, giving a red-brown tint to aged samples.

## Production

Aniline is mainly produced in industry in two steps from benzene. First, benzene is nitrated using a concentrated mixture of nitric acid and sulfuric acid at 50 to 60°C, which gives nitrobenzene. In the second step, the nitrobenzene is hydrogenated, typically at 200-300 °C in presence of various metal catalysts:

$$C_6H_5NO_2 + 3\ H_2 \rightarrow C_6H_5NH_2 + 2\ H_2O$$

Originally, the reduction was effected with a mixture of ferrous chloride and iron metal via the Bechamp reduction.

As an alternative, aniline is also prepared from phenol and ammonia, the phenol being derived from the cumene process.

In commerce, three brands of aniline are distinguished: aniline oil for blue, which is pure aniline; aniline oil for red, a mixture of equimolecular quantities of aniline and ortho- and para-toluidines; and aniline oil for safranine, which contains aniline and ortho-toluidine, and is obtained from the distillate of the fuchsine fusion.

## Related Aniline Derivatives

Many derivatives of aniline can be prepared in similar fashion from nitrated aromatic compounds. Nitration followed by reduction of toluene affords toluidines. Nitration of chlorobenzene and related derivatives and reduction of the nitration products gives aniline derivatives, e.g. 4-chloroaniline.

## Reactions

The chemistry of aniline is extremely rich because the compound has been cheaply available for many years. Below are some classes of its reactions.

## Oxidation

The oxidation of aniline has been heavily investigated, and can result in reactions localized at nitrogen or more commonly results in the formation of new C-N bonds. In alkaline solution, azobenzene results, whereas arsenic acid produces the violet-coloring matter violaniline.

Chromic acid converts it into quinone, whereas chlorates, in the presence of certain metallic salts (especially of vanadium), give aniline black. Hydrochloric acid and potassium chlorate give chloranil. Potassium permanganate in neutral solution oxidizes it to nitrobenzene, in alkaline solution to azobenzene, ammonia and oxalic acid, in acid solution to aniline black. Hypochlorous acid gives 4-aminophenol and para-amino diphenylamine. Oxidation with persulfate affords a variety of polyanilines compounds. These polymers exhibit rich redox and acid-base properties.

### *Electrophilic Reactions at Carbon*

Like phenols, aniline derivatives are highly susceptible to electrophilic substitution reactions. Its high reactivity reflects that it is an enamine, which enhances the electron-donating ability of the ring. For example, reaction of aniline with sulfuric acid at 180 °C produces sulfanilic acid, $H_2NC_6H_4SO_3H$, which can be converted to sulfanilamide. Sulfanilamide is one of the sulfa drugs, which were widely used as antibacterials in the early 20th century. The largest scale industrial reaction of aniline involves its alkylation with formaldehyde:

$$2\ C_6H_5NH_2 + CH_2O \rightarrow CH_2(C_6H_4NH_2)_2 + H_2O$$

The resulting diamine is the precursor to 4,4'-MDI and related diisocyanates.

## Reactions at Nitrogen

### *Basicity*

Aniline is a weak base. Aromatic amines such as aniline are, in general, much weaker bases than aliphatic amines because of the electron-withdrawing effect of the phenyl group. Aniline reacts with strong acids to form anilinium (or phenylammonium) ion ($C_6H_5$-$NH_3^+$). The sulfate forms white plates. Although aniline is weakly basic, it precipitates zinc, aluminium, and ferric salts, and, on warming, expels ammonia from its salts. The weak basicity is due to a negative inductive effect as the lone pair on the nitrogen is partially delocalized into the pi system of the benzene ring.

### *Acylation*

Aniline reacts with carboxylic acids or more readily with acyl chlorides such as acetyl chloride to give amides. The amides formed from aniline are sometimes called anilides, for example $CH_3$-CO-NH-$C_6H_5$ is acetanilide. Antifebrin (acetanilide), an anti-pyretic and analgesic, is obtained by the reaction of acetic acid and aniline.

### N-*Alkylation*

N-methylation of aniline with methanol at elevated temperatures over acid catalsts gives *N*-methylaniline and dimethylaniline:

$$C_6H_5NH_2 + 2\ CH_3OH \rightarrow C_6H_4N(CH_3)_2 + H_2O$$

Methyl and dimethylaniline are colourless liquids with b.p. of 193-195 °C and 192 °C, respectively. These derivatives are of importance in the colour industry. Aniline combines directly with alkyl iodides to form secondary and tertiary amines.

### *Carbon Disulfide Derivatives*

Boiled with carbon disulfide, it gives sulfocarbanilide (diphenylthiourea) ($CS(NHC_6H_5)_2$), which may be decomposed into phenyl isothiocyanate($C_6H_5CNS$), and triphenyl guanidine ($C_6H_5N{=}C(NHC_6H_5)_2$).

### *Diazotization*

Aniline and its ring-substituted derivatives react with nitrous acid to form diazonium salts. Through these intermediates, aniline can be conveniently converted to -OH, -CN, or a halide via Sandmeyer reactions. This diazonium salt can also be reacted with NaNO2 and phenol which produces a dye which is benzeneazophenol, this process is called coupling.

### *Other Reactions*

It reacts with nitrobenzene to produce phenazine in the Wohl-Aue reaction. Hydrogenation gives cyclohexylamine.

Being a standard reagent in laboratories, aniline is used for many niche reactions. Its acetate is used in the Aniline acetate test for carbohydrates, identifying pentoses by conversion to furfural. It is used to stain neural RNA blue in the Nissl stain.

### *Uses*

The largest application of aniline is for the preparation of methylene diphenyl diisocyanate (MDI). The majority of aniline serves this market. Other uses include rubber processing chemicals (9%), herbicides (2%), and dyes and pigments (2%). As additives to rubber, aniline derivatives such as phenylenediamine and diphenylamine, are antioxidants. Illustrative of the drugs prepared from aniline is paracetamol (acetaminophen, Tylenol). The principal use of aniline in the dye industry is as a precursor to indigo, the blue of blue jeans.

Aniline is also used at a smaller scale in the production of the intrinsically conducting polymer polyaniline.

### History

Aniline was first isolated from the destructive distillation of indigo in 1826 by Otto Unverdorben, who named it crystalline. In 1834, Friedlieb Runge isolated from coal tar a substance that produced a beautiful blue colour on treatment with chloride of lime, which he named kyanol or cyanol.

In 1841, C. J. Fritzsche showed that, by treating indigo with caustic potash, it yielded an oil, which he named aniline from the name of one of the indigo-yielding plants, Añil (*Indigofera suffruticosa*, syn. *I. anil*, ultimately from Sanskrit "nîla", dark-blue). About the same time N. N. Zinin found that, on reducing nitrobenzene, a base was formed, which he named benzidam. August Wilhelm von Hofmann investigated these variously-prepared substances, and proved them to be identical (1855), and thenceforth they took their place as one body, under the name aniline or phenylamine.

The great commercial value of aniline was due to the readiness with which it yields, directly or indirectly, dyestuffs. The discovery of mauve in 1856 by William Henry Perkin was the first of a series of an enormous range of dyestuffs, such as fuchsine, safranine and induline.

Its first industrial-scale use was in the manufacture of mauveine, a purple dye discovered in 1856 by Hofmann's student William Henry Perkin. At the time of mauveine's discovery, aniline was an expensive laboratory compound, but it was soon prepared "by the ton" using a process previously discovered by Antoine Béchamp. The synthetic dye industry grew rapidly as new aniline-based dyes were discovered in the late 1850s and 1860s, an echo of the importance of the compound being found in the name of one of the world's largest chemical companies, BASF, originally the Badische Anilin- und Soda-Fabrik.

### Toxicology

Aniline is toxic by inhalation of the vapour. The IARC lists it in Group 3 (*not classifiable as to its carcinogenicity to humans*) due to the limited and contradictory data available. The early manufacture of aniline resulted in increased incidents of bladder cancer, but these effects are now attributed to naphthylamines, not anilines.

## Methylene Diphenyl Diisocyanate

Methylene diphenyl diisocyanate, most often abbreviated as MDI, is an aromatic diisocyanate. It exists in three isomers, 2,2'-MDI, 2,4'-MDI, and 4,4'-MDI, but the 4,4' isomer is most widely used. This isomer is also known as Pure MDI. MDI reacts with polyols in the manufacture of polyurethane. It is the most produced diisocyanate, accounting for 61.3% of the global market in the year 2000.

### Production

Total world production of MDI and polymeric MDI is over 2 million tonnes per year (Mt/a). Major producers include BASF, Bayer, BorsodChem, Dow, Huntsman, Mitsui, Nippon Polyurethane Industry and Yantai Wanhua. All major producers of MDI are members of the International Isocyanate Institute, whose aim is the promotion of the safe handling of MDI and TDI in the workplace, community and environment.

MDI is prepared by "phosgenation" of a diamine precursor. These diamines are treated with phosgene to form an MDI. The isomer ratio is determined by the isomeric composition of the diamine. Distillation of the MDI mixture give Polymeric MDI (a mixture of oligomeric polyisocyanates) and an MDI isomer mixture which has a low 2,4' isomer content. Further purification entails fractionation of the MDI isomer mixture.

### Reactivity of the Isocyanate Group

The positions of the isocyanate groups influences their reactivity. In 4,4'-MDI, the two isocyanate groups are equivalent but in 2,4'-MDI the two groups display highly differing reactivities. The group at the 4-position is approximately four times more reactive than the group at the 2-position.

### Applications

The major application of 4,4'-MDI is the production of rigid polyurethane. Typically, one tonne of polyurethane foam needs 0.616 tonne of MDI and 0.386 tonne of polyol, with 0.054 tonne pentane as a blowing agent. These rigid polyurethane foams are good thermal insulators and used in nearly all freezers and refrigerators worldwide, as well as buildings. Typical polyols used are polyethylene adipate (a polyester) and poly(tetramethylene ether) glycol (a polyether).

4,4'-MDI is also used as an industrial strength adhesive, which is available to end consumers as various high-strength bottled glue preparations.

### *Safety*

MDI is the least hazardous of the commonly available isocyanates but is not benign. Its very low vapour pressure reduces its hazards during handling compared to the other major isocyanates (TDI, HDI). However, it, like the other isocyanates, is an allergen and sensitizer. Persons developing sensitivity to isocyanates may have dangerous systemic reactions to extremely small exposures, including respiratory failure. Handling MDI requires strict engineering controls and personal protective equipment. Compared to other organic cyanates, MDI has a relatively low human toxicity. It is potentially violently reactive material toward water and other nucleophiles.

## Polyurethane

A polyurethane (PUR and PU) is any polymer composed of a chain of organic units joined by carbamate (urethane) links. Polyurethane polymers are formed through step-growth polymerization, by reacting a monomer (with at least two isocyanate functional groups) with another monomer (with at least two hydroxyl or alcohol groups) in the presence of a catalyst.

Polyurethanes are applied to the manufacture of flexible, high-resilience foam seating; rigid foam insulation panels; microcellular foam seals and gaskets; durable elastomeric wheels and tires; automotive suspension bushings; electrical potting compounds; high performance adhesives; surface coatings and surface sealants; synthetic fibres (e.g. Spandex); carpet underlay; and hard-plastic parts (i.e. for electronic instruments). Moreover, polyurethane products often are inaccurately called “urethanes”, but must not be confused with urethane proper (ethyl carbamate), because polyurethanes neither contain nor are produced from ethyl carbamate.

## History

The pioneering work on polyurethane polymers was conducted by Otto Bayer and his coworkers in 1937 at the laboratories of I.G. Farben in Leverkusen, Germany. They recognized that using the polyaddition principle to produce polyurethanes from liquid diisocyanates and liquid polyether or polyester diols seemed to point to special opportunities, especially when compared to already existing plastics that were made by polymerizing olefins, or by polycondensation. The new monomer combination also circumvented existing patents obtained by Wallace Carothers on polyesters. Initially, work focused on the production of fibres and flexible foams. With development constrained by World War II (when PUs were applied on a limited scale as aircraft coating), it was not until 1952 that polyisocyanates became commercially available. Commercial production of flexible polyurethane foam began in 1954, based on toluene diisocyanate (TDI) and polyester polyols. The invention of these foams (initially called *imitation swiss cheese* by the inventors) was thanks to water accidentally introduced in the reaction mix. These materials were also used to produce rigid foams, gum rubber, and elastomers. Linear fibres were produced from hexamethylene diisocyanate (HDI) and 1,4-butanediol (BDO).

The first commercially available polyether polyol, poly(tetramethylene ether) glycol, was introduced by DuPont in 1956 by polymerizing tetrahydrofuran. Less expensive polyalkylene glycols were introduced by BASF and Dow Chemical in 1957. Polyether polyols offered technical and commercial advantages such as low cost, ease of handling, and better hydrolytic stability over polyester polyols and quickly replaced them in the manufacture of polyurethane goods. Other PU pioneers were Union Carbide and Mobay, a U.S. Monsanto/Bayer joint venture. In 1960 more than 45,000 metric tons of flexible polyurethane foams were produced. As the decade progressed, the availability of chlorofluoroalkane blowing agents, inexpensive polyether polyols, and methylene diphenyl diisocyanate (MDI) heralded the development and use of polyurethane rigid foams as high performance insulation materials. Rigid foams based on polymeric MDI (PMDI) offered better thermal stability and combustion characteristics than those based on TDI. In 1967, urethane modified polyisocyanurate rigid foams were introduced, offering even better thermal stability and flammability resistance compared to low-density insulation products. During the 1960s, automotive interior safety components such as instrument and door panels were produced by back-filling thermoplastic skins with semi-rigid foam.

In 1969, Bayer exhibited an all plastic car in Düsseldorf, Germany. Parts of this car were manufactured using a new process called RIM, Reaction Injection Molding. RIM technology uses high pressure impingement of liquid reactive components followed by the rapid flow of the reaction mixture into a mold cavity. Large parts, such as automotive fascia and body panels, can be molded in this manner. Polyurethane RIM evolved into a number of different products and processes. Using diamine chain extenders and trimerization technology gave poly(urethane urea), poly(urethane isocyanurate), and polyurea RIM. The addition of fillers, such as milled glass, mica, and processed mineral fibres gave rise to reinforced RIM (RRIM), which provided improvements in flexural modulus (stiffness), reduction in coefficient of thermal expansion and thermal stability. This technology allowed production of the first plastic-body automobile in the United States, the Pontiac Fiero, in 1983. Further increases in flexural modulus were obtained by incorporating pre-placed glass mats into the RIM mold cavity, also known broadly as resin injection molding (a process technology that also includes thermosetting polyester resins, epoxide resins, etc.) or more specifically for PUR systems as SRIM, or structural RIM.

Starting in the early 1980s, water-blown microcellular flexible foams were used to mold gaskets for panel and radial seal air filters in the automotive industry. Since then, increasing energy prices and the pressures to eliminate PVC plastisol from automotive applications have greatly increased market share. Costlier raw materials are offset by a significant decrease in part weight and in some cases, the elimination of metal end caps and filter housings. Highly filled polyurethane elastomers, and more recently unfilled polyurethane foams are now used in high temperature oil filter applications.

Polyurethane foam (including foam rubber) is often made by adding small amounts of volatile materials, so-called blowing agents, to the reaction mixture. These volatile chemicals yield important performance characteristics, primarily density reduction, cushioning/energy absorption and thermal insulation. In the early 1990s, because of their impact on ozone depletion, the Montreal Protocol led to the greatly reduced use of many chlorine-containing blowing agents, such as trichlorofluoromethane (CFC-11). Other haloalkanes, such as the hydrochlorofluorocarbon 1,1-dichloro-1-fluoroethane (HCFC-141b), were used as interim replacements until their phase out under the IPPC directive on greenhouse gases in 1994 and by the Volatile Organic Compounds (VOC) directive of the EU in 1997. By the late 1990s, the use of blowing agents such as carbon dioxide, pentane, 1,1,1,2-tetrafluoroethane (HFC-134a) and 1,1,1,3,3-pentafluoropropane (HFC-245fa) became more widespread in North America and the EU, although chlorinated blowing agents remained in use in many developing countries.

Building on existing polyurethane spray coating technology and polyetheramine chemistry, extensive development of two-component polyurea spray elastomers took place in the 1990s. Their fast reactivity and relative insensitivity to moisture make them useful coatings for large surface area projects, such as secondary containment, manhole and tunnel coatings, and tank liners. Excellent adhesion to concrete and steel is obtained with the proper surface treatment and primer. During the same period, new two-component polyurethane and hybrid polyurethane-polyurea elastomer technology was used to enter the marketplace of spray-in-place load bed liners and military marine applications for the U.S. Navy. Even a one-part polyurethane is specified as high durability deck coatings under MIL-PRF-32171 for the US Navy. This technique for coating creates a durable, abrasion resistant composite with the metal substrate, and eliminates corrosion and brittleness associated with drop-in thermoplastic bed liners.

The potential for polyols derived from vegetable oils to replace petrochemical-based polyols began garnering attention beginning around 2004, partly due to the rising costs of petrochemical feedstocks and partially due to an enhanced public desire for environmentally friendly green products. One of the most vocal supporters of these polyurethanes made using natural oil polyols is the Ford Motor Company.

## Chemistry

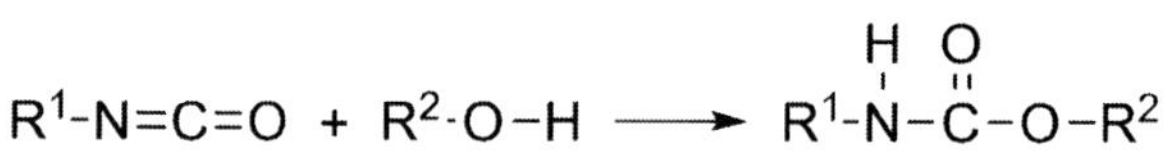

*Generalized polyurethane reaction*

Polyurethanes are in the class of compounds called reaction polymers, which include epoxies, unsaturated polyesters, and phenolics. A urethane linkage is produced by reacting an isocyanate group, -N=C=O with a hydroxyl (alcohol) group, -OH. Polyurethanes are produced by the polyaddition reaction of a polyisocyanate with a polyalcohol (polyol) in the presence of a catalyst and other additives. In this case, a polyisocyanate is a molecule with two or more isocyanate functional groups, R-(N=C=O)$_{n\ e"\ 2}$ and a polyol is a molecule with two or more hydroxyl functional groups, R'-(OH)$_{n\ e"\ 2}$. The reaction product is a polymer containing the urethane linkage, -RNHCOOR'-. Isocyanates will react with any molecule that contains an active hydrogen.

Importantly, isocyanates react with water to form a urea linkage and carbon dioxide gas; they also react with polyetheramines to form polyureas. Commercially, polyurethanes are produced by reacting a liquid isocyanate with a liquid blend of polyols, catalyst, and other additives. These two components are referred to as a polyurethane system, or simply a system. The isocyanate is commonly referred to in North America as the 'A-side' or just the 'iso'. The blend of polyols and other additives is commonly referred to as the 'B-side' or as the 'poly'. This mixture might also be called a 'resin' or 'resin blend'. In Europe the meanings for 'A-side' and 'B-side' are reversed. Resin blend additives may include chain extenders, cross linkers, surfactants, flame retardants, blowing agents, pigments, and fillers.

The first essential component of a polyurethane polymer is the isocyanate. Molecules that contain two isocyanate groups are called diisocyanates. These molecules are also referred to as monomers or monomer units, since they themselves are used to produce polymeric isocyanates that contain three or more isocyanate functional groups. Isocyanates can be classed as aromatic, such as diphenylmethane diisocyanate (MDI) or toluene diisocyanate (TDI); or aliphatic, such as hexamethylene diisocyanate (HDI) or isophorone diisocyanate (IPDI). An example of a polymeric isocyanate is polymeric diphenylmethane diisocyanate, which is a blend of molecules with two-, three-, and four- or more isocyanate groups, with an average functionality of 2.7. Isocyanates can be further modified by partially reacting them with a polyol to form a prepolymer. A quasi-prepolymer is formed when the stoichiometric ratio of isocyanate to hydroxyl groups is greater than 2:1. A true prepolymer is formed when the stoichiometric ratio is equal to 2:1. Important characteristics of isocyanates are their molecular backbone, % NCO content, functionality, and viscosity.

The second essential component of a polyurethane polymer is the polyol. Molecules that contain two hydroxyl groups are called diols, those with three hydroxyl groups are called triols, et cetera. In practice, polyols are distinguished from short chain or low-molecular weight glycol chain extenders and cross linkers such as ethylene glycol (EG), 1,4-butanediol (BDO), diethylene glycol (DEG), glycerine, and trimethylolpropane (TMP).

Polyols are polymers in their own right. They are formed by base-catalysed addition of propylene oxide (PO), ethylene oxide (EO) onto a hydroxyl or amine containing initiator, or by polyesterification of a di-acid, such as adipic acid, with glycols, such as ethylene glycol or dipropylene glycol (DPG). Polyols extended with PO or EO are polyether polyols. Polyols formed by polyesterification are polyester polyols. The choice of initiator, extender, and molecular weight of the polyol greatly affect its physical state, and the physical properties of the polyurethane polymer. Important characteristics of polyols are their molecular backbone, initiator, molecular weight, % primary hydroxyl groups, functionality, and viscosity.

*PU reaction mechanism catalysed by a tertiary amine*

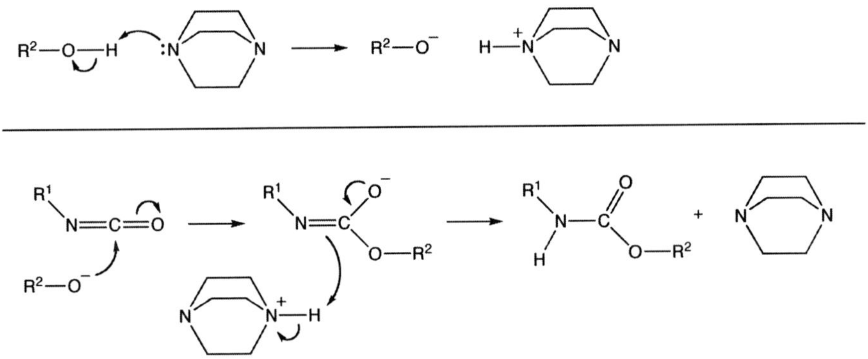

*carbon dioxide gas formed by reacting water and isocyanate*

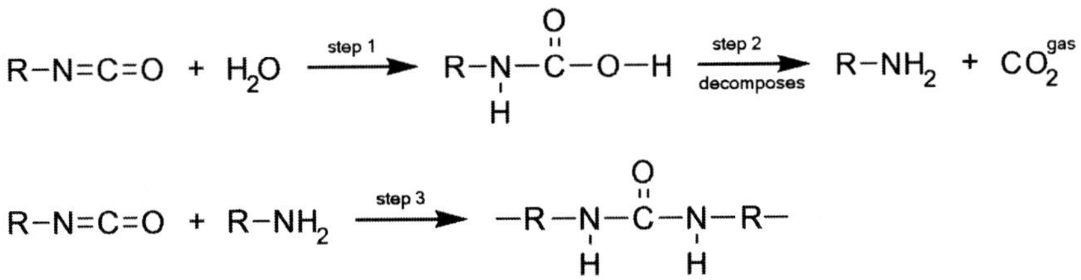

The polymerization reaction is catalysed by tertiary amines, such as dimethylcyclohexylamine, and organometallic compounds, such as dibutyltin dilaurate or bismuth octanoate. Furthermore, catalysts can be chosen based on whether they favour the urethane (gel) reaction, such as 1,4-diazabicyclo[2.2.2]octane (also called DABCO or TEDA), or the urea (blow) reaction, such as bis-(2-dimethylaminoethyl)ether, or specifically drive the isocyanate trimerization reaction, such as potassium octoate.

. One of the most desirable attributes of polyurethanes is their ability to be turned into foam. Blowing agents such as water, certain halocarbons such as HFC-245fa (1,1,1,3,3-pentafluoropropane) and HFC-134a (1,1,1,2-tetrafluoroethane), and hydrocarbons such as n-pentane, can be incorporated into the poly side or added as an auxiliary stream. Water reacts with the isocyanate to create carbon dioxide gas, which fills and expands cells created during the mixing process. The reaction is a three step process.

A water molecule reacts with an isocyanate group to form a carbamic acid. Carbamic acids are unstable, and decompose forming carbon dioxide and an amine. The amine reacts with more isocyanate to give a substituted urea. Water has a very low molecular weight, so even though the weight percent of water may be small, the molar proportion of water may be high and considerable amounts of urea produced. The urea is not very soluble in the reaction mixture and tends to form separate "hard segment" phases consisting mostly of polyurea.

The concentration and organization of these polyurea phases can have a significant impact on the properties of the polyurethane foam. Halocarbons and hydrocarbons are chosen such that they have boiling points at or near room temperature. Since the polymerization reaction is exothermic, these blowing agents volatilize into a gas during the reaction process. They fill and expand the cellular polymer matrix, creating a foam. It is important to know that the blowing gas does not create the cells of a foam. Rather, foam cells are a result of blowing gas diffusing into bubbles that are nucleated or stirred into the system at the time of mixing. In fact, high-density microcellular foams can be formed without the addition of blowing agents by mechanically frothing or nucleating the polyol component prior to use.

Surfactants are used to modify the characteristics of the polymer during the foaming process. They are used to emulsify the liquid components, regulate cell size, and stabilize the cell structure to prevent collapse and surface defects. Rigid foam surfactants are designed to produce very fine cells and a very high closed cell content. Flexible foam surfactants are designed to stabilize the reaction mass while at the same time maximizing open cell content to prevent the foam from shrinking. The need for surfactant can be affected by choice of isocyanate, polyol, component compatibility, system reactivity, process conditions and equipment, tooling, part shape, and shot weight.

Though the properties of the polyurethane are determined mainly by the choice of polyol, the diisocyanate exerts some influence, and must be suited to the application. The cure rate is influenced by the functional group reactivity and the number of functional isocyanate groups. The mechanical properties are influenced by the functionality and the molecular shape. The choice of diisocyanate also affects the stability of the polyurethane upon exposure to light. Polyurethanes made with aromatic diisocyanates yellow with exposure to light, whereas those made with aliphatic diisocyanates are stable.

Softer, elastic, and more flexible polyurethanes result when linear difunctional polyethylene glycol segments, commonly called polyether polyols, are used to create the urethane links. This strategy is used to make spandex elastomeric fibres and soft rubber parts, as well as foam rubber. More rigid products result if polyfunctional polyols are used, as these create a three-dimensional cross-linked structure which, again, can be in the form of a low-density foam.

An even more rigid foam can be made with the use of specialty trimerization catalysts which create cyclic structures within the foam matrix, giving a harder, more thermally stable structure, designated as polyisocyanurate foams. Such properties are desired in rigid foam products used in the construction sector.

Careful control of viscoelastic properties — by modifying the catalysts and polyols used —can lead to memory foam, which is much softer at skin temperature than at room temperature.

There are then two main foam variants: one in which most of the foam bubbles (cells) remain closed, and the gas(es) remains trapped, the other being systems which have mostly open cells, resulting after a critical stage in the foam-making process (if cells did not form, or became open too soon, foam would not be created). This is a vitally important process: if the flexible foams have closed cells, their softness is severely compromised, they become pneumatic in feel, rather than soft; so, generally speaking, flexible foams are required to be open-celled.

The opposite is the case with most rigid foams. Here, retention of the cell gas is desired since this gas (especially the fluorocarbons referred to above) gives the foams their key characteristic: high thermal insulation performance. A third foam variant, called microcellular foam, yields the tough elastomeric materials typically experienced in the coverings of car steering wheels and other interior automotive components.

### Raw Materials

For the manufacture of polyurethane polymers, two groups of at least bifunctional substances are needed as reactants; compounds with isocyanate groups, and compounds with active hydrogen atoms. The physical and chemical character, structure, and molecular size of these compounds influence the polymerization reaction, as well as ease of processing and final physical properties of the finished polyurethane. In addition, additive such as catalysts, surfactants, blowing agents, cross linkers, flame retardants, light stabilizers, and fillers are used to control and modify the reaction process and performance characteristics of the polymer.

### Isocyanates

Isocyanates with two or more functional groups are required for the formation of polyurethane polymers. Volume wise, aromatic isocyanates account for the vast majority of global diisocyanate production. Aliphatic and cycloaliphatic isocyanates are also important building blocks for polyurethane materials, but in much smaller volumes. There are a number of reasons for this. First, the aromatically linked isocyanate group is much more reactive than the aliphatic one. Second, aromatic isocyanates are more economical to use. Aliphatic isocyanates are used only if special properties are required for the final product. For example, light stable coatings and elastomers can only be obtained with aliphatic isocyanates. Even within the same class of isocyanates, there is a significant difference in reactivity of the functional groups based on steric hindrance. In the case of 2,4-toluene diisocyanate, the isocyanate group in the para position to the methyl group is much more reactive than the isocyanate group in the ortho position.

Phosgenation of corresponding amines is the main technical process for the manufacture of isocyanates. The amine raw materials are generally manufactured by the hydrogenation of corresponding nitro compounds. For example, toluenediamine (TDA) is manufactured from dinitrotoluene, which then converted to toluene diisocyanate (TDI). Diamino diphenylmethane or methylenedianiline (MDA) is manufactured from nitrobenzene via aniline, which is then converted to diphenylmethane diisocyanate (MDI).

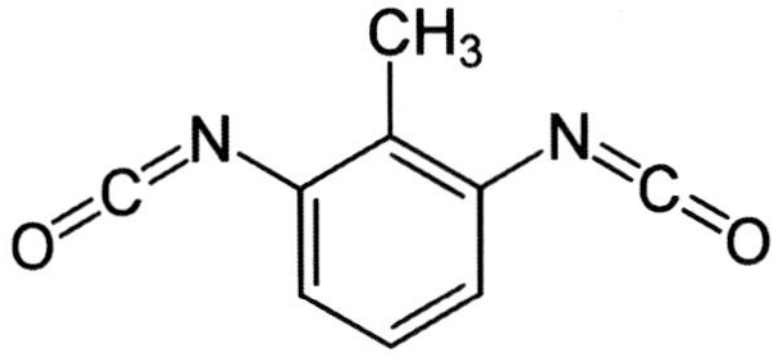


*2,6-Diisocyanatotoluene*

The two most important commercial, aromatic isocyanates are toluene diisocyanate (TDI) and diphenylmethane diisocyanate (MDI). TDI consists of a mixture of the 2,4- and 2,6-diisocyanatotoluene isomers. The most important product is TDI-80 (TD-80), consisting of 80% of the 2,4-isomer and 20% of the 2,6-isomer. This blend is used extensively in the manufacture of polyurethane flexible slabstock and molded foam. TDI, and especially crude TDI and TDI/MDI blends can be used in rigid foam applications, but have been supplanted by polymeric MDI. TDI-polyether and TDI-polyester prepolymers are used in high performance coating and elastomer applications. Prepolymers are available that have been vacuum stripped of TDI monomer, which greatly reduces their toxicity.

Diphenylmethane diisocyanate (MDI) has three isomers, 4,4'-MDI, 2,4'-MDI, and 2,2'-MDI, and is also polymerized to provide oligomers of functionality three and higher.

4,4

2,4'

2,2'

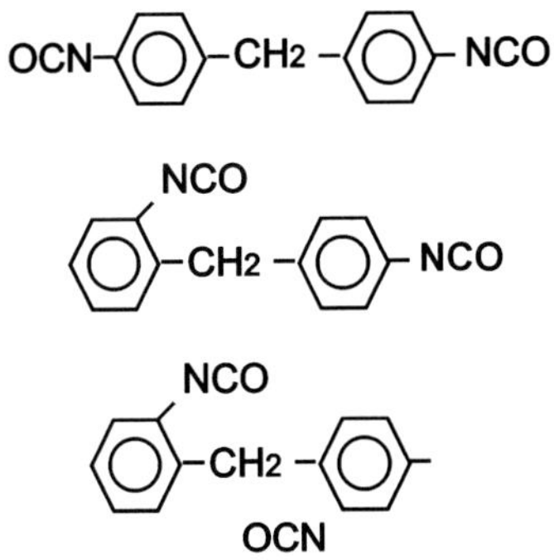


Pure MDI's

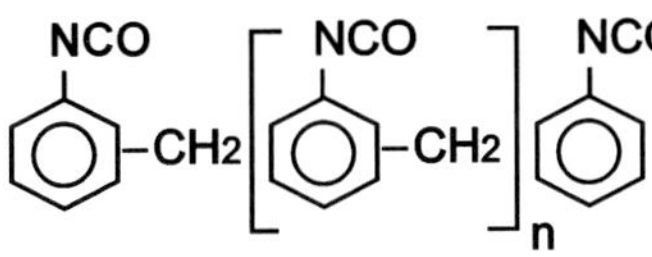


Polymeric MDI's

Only the 4,4'-MDI monomer is sold commercially as a single isomer. It is provided either as a frozen solid or flake, or in molten form, and is used to manufacture high performance prepolymers. Monomer blends, consisting of approximately 50% of the 4,4'-isomer and 50% of the 2,4'-

isomer, are liquid at room temperature and are used to manufacture prepolymers for polyurea spray elastomer applications. 4,4'-MDI blends containing MDI uretonimine, carbodiimide, and allophonate moieties are also liquid at room temperature, and are used in the manufacture of integral skin and microcellular foams. 4,4'-MDI-glycol prepolymers offer increased mechanical properties in the same applications, but are prone to freezing at temperatures below 20 °C.

Polymeric MDI (PMDI) is used in rigid pour-in-place, spray foam, and molded foam applications. Polymeric MDI that contains a very high portion of high-functionality oligomers is used to manufacture polyurethane and polyisocyanurate rigid insulation boardstock. Modified PMDI, which contains high levels of MDI monomer, is used in the production of polyurethane flexible molded and microcellular foams. The relative percentage of the 4,4'- and 2,4'- isomers is adjusted to change the reactivity and storage stability of the isocyanate blend, as well as the firmness and other physical properties of the finished goods. Other aromatic isocyanate include p-phenylene diisocyanate (PPDI), naphthalene diisocyanate (NDI), and o-tolidine diisocyanate (TODI).

The most important aliphatic and cycloaliphatic isocyanates are 1,6-hexamethylene diisocyanate (HDI), 1-isocyanato-3-isocyanatomethyl-3,5,5-trimethyl-cyclohexane (isophorone diisocyanate, IPDI), and 4,4'-diisocyanato dicyclohexylmethane ($H_{12}$MDI or hydrogenated MDI). They are used to produce light stable, non-yellowing polyurethane coatings and elastomers. Because of their volatility and toxicity, aliphatic isocyanate monomers are converted into prepolymers, biurets, dimers, and trimers for commercial use. HDI adducts are used extensively for weather and abrasion resistant coatings and lacquers. IPDI is used in the manufacture of coatings, elastomeric adhesives and sealants. $H_{12}$MDI prepolymers are used to produce high performance coatings and elastomers with optical clarity and hydrolysis resistance. Other aliphatic isocyanates include cyclohexane diisocyanate (CHDI), tetramethylxylene diisocyanate (TMXDI), and 1,3-bis(isocyanatomethyl)cyclohexane ($H_6$XDI).

### *Polyols*

Polyols are higher molecular weight materials manufactured from an initiator and monomeric building blocks. They are most easily classified as polyether polyols, which are made by the reaction of epoxides (oxiranes) with an active hydrogen containing starter compounds, or polyester polyols, which are made by the polycondensation of multifunctional carboxylic acids and hydroxyl compounds. They can be further classified according to their end use as flexible or rigid polyols, depending on the functionality of the initiator and their molecular weight. Taking into account functionality, flexible polyols have molecular weights from 2,000 to 10,000 (OH# from 18 to 56). Rigid polyols have molecular weights from 250 to 700 (OH# from 300 to 700). Polyols with molecular weights from 700 to 2,000 (OH# 60 to 280) are used to add stiffness or flexibility to base systems, as well as increase solubility of low molecular weight glycols in high molecular weight polyols.

Polyether polyols come in a wide variety of grades based on their end use, but are all constructed in a similar manner. Polyols for flexible applications use low functionality initiators such as dipropylene glycol (f=2), glycerine (f=3) or a sorbitol/water solution (f=2.75). Polyols for rigid applications use high functionality initiators such sucrose (f=8), sorbitol (f=6), toluenediamine (f=4), and Mannich bases (f=4). Propylene oxide is then added to the initiators until the desired molecular weight is achieved. Polyols extended with propylene oxide are terminated with secondary hydroxyl groups. In order to change the compatibility, rheological properties, and reactivity of a polyol, ethylene oxide is used as a co-reactant to create random or mixed block heteropolymers. Polyols capped with ethylene oxide contain a high percentage of primary hydroxyl groups, which are more reactive than secondary hydroxyl groups. Because of their high viscosity (470 OH# sucrose polyol, 33 Pa·s at 25 °C), carbohydrate initiated polyols often use glycerine or diethylene glycol as a co-initiate in order to lower the viscosity to ease handling and processing (490 OH# sucrose-glycerine polyol, 5.5 Pa·s at 25 °C). Graft polyols (also called filled polyols or polymer polyols) contain finely dispersed styrene-acrylonitrile, acrylonitrile, or polyurea (PHD) polymer solids chemically grafted to a high molecular weight polyether backbone. They are used to increase the load-bearing properties of low-density high-resiliency (HR) foam, as well as add toughness to microcellular foams and cast elastomers. PHD polyols are also used to modify the combustion properties of HR flexible foam. Solids content ranges from 14% to 50%, with 22% and 43% being typical. Initiators such as ethylenediamine and triethanolamine are used to make low molecular weight rigid foam polyols that have built-in catalytic activity due to the presence of nitrogen atoms in the backbone. They are used to increase system reactivity and physical property build, and to reduce the friability of rigid foam molded parts. A special class of polyether polyols, poly(tetramethylene ether) glycols are made by polymerizing tetrahydrofuran. They are used in high performance coating and elastomer applications.

Polyester polyols fall into two distinct categories according to composition and application. Conventional polyester polyols are based on virgin raw materials and are manufactured by the direct polyesterification of high-purity diacids and glycols, such as adipic acid and 1,4-butanediol. They are distinguished by the choice of monomers, molecular weight, and degree of branching. While costly and difficult to handle because of their high viscosity, they offer physical properties not obtainable with polyether polyols, including superior solvent, abrasion, and cut resistance. Other polyester polyols are based on reclaimed raw materials. They are manufactured by transesterification (glycolysis) of recycled poly(ethyleneterephthalate) (PET) or dimethylterephthalate (DMT) distillation bottoms with glycols such as diethylene glycol. These low molecular weight, aromatic polyester polyols are used in the manufacture of rigid foam, and bring low cost and excellent flammability characteristics to polyisocyanurate (PIR) boardstock and polyurethane spray foam insulation.

Many polyols are polydispersive materials, being blends of two or more polyols each of specific molecular weights, to give intermediate molecular weight materials. It is not unusual to find blends of polyether and polyester polyols, to give specific compromises in properties.

Specialty polyols include polycarbonate polyols, polycaprolactone polyols, polybutadiene polyols, and polysulfide polyols. The materials are used in elastomer, sealant, and adhesive applications that require superior weatherability, and resistance to chemical and environmental attack. Natural oil polyols derived from castor oil and other vegetable oils are used to make elastomers, flexible bunstock, and flexible molded foam.

Copolymerizing chlorotrifluoroethylene or tetrafluoroethylene with vinyl ethers containing hydroxyalkyl vinyl ether produces fluorinated (FEVE) polyols. Two component fluorinated polyurethane prepared by reacting FEVE fluorinated polyols with polyisocyanate have been applied for make ambient cure paint/coating. Since fluorinated polyurethanes contain high percentage of fluorine-carbon bond which is the strongest bond among all chemical bonds. Fluorinated polyurethanes have excellent resistance to UV, acids, alkali, salts, chemicals, solvents, weathering, corrosion, fugi and microbial attack. These have become the first choice for high performance coating/ paints.

### *Chain Extenders and Cross Linkers*

Chain extenders (f=2) and cross linkers (f=3 or greater) are low molecular weight hydroxyl and amine terminated compounds that play an important role in the polymer morphology of polyurethane fibres, elastomers, adhesives, and certain integral skin and microcellular foams. The elastomeric properties of these materials are derived from the phase separation of the hard and soft copolymer segments of the polymer, such that the urethane hard segment domains serve as cross-links between the amorphous polyether (or polyester) soft segment domains.

This phase separation occurs because the mainly non-polar, low melting soft segments are incompatible with the polar, high melting hard segments. The soft segments, which are formed from high molecular weight polyols, are mobile and are normally present in coiled formation, while the hard segments, which are formed from the isocyanate and chain extenders, are stiff and immobile. Because the hard segments are covalently coupled to the soft segments, they inhibit plastic flow of the polymer chains, thus creating elastomeric resiliency. Upon mechanical deformation, a portion of the soft segments are stressed by uncoiling, and the hard segments become aligned in the stress direction.

This reorientation of the hard segments and consequent powerful hydrogen bonding contributes to high tensile strength, elongation, and tear resistance values. The choice of chain extender also determines flexural, heat, and chemical resistance properties. The most important chain extenders are ethylene glycol, 1,4-butanediol (1,4-BDO or BDO), 1,6-hexanediol, cyclohexane dimethanol and hydroquinone bis(2-hydroxyethyl) ether (HQEE). All of these glycols form polyurethanes that phase separate well and form well defined hard segment domains, and are melt processable. They are all suitable for thermoplastic polyurethanes with the exception of ethylene glycol, since its derived bis-phenyl urethane undergoes unfavorable degradation at high hard segment levels. Diethanolamine and triethanolamine are used in flex molded foams to build firmness and add catalytic activity. Diethyltoluenediamine is used extensively in RIM, and in polyurethane and polyurea elastomer formulations.

*Table 1 : Table of chain extenders and cross linkers hydroxyl compounds – difunctional molecules*

| | ***MW*** | ***s.g.*** | ***m.p. °C*** | ***b.p. °C*** |
|---|---|---|---|---|
| ethylene glycol | 62.1 | 1.110 | -13.4 | 197.4 |
| diethylene glycol | 106.1 | 1.111 | -8.7 | 245.5 |
| triethylene glycol | 150.2 | 1.120 | -7.2 | 287.8 |
| tetraethylene glycol | 194.2 | 1.123 | -9.4 | 325.6 |
| propylene glycol | 76.1 | 1.032 | supercools | 187.4 |
| dipropylene glycol | 134.2 | 1.022 | supercools | 232.2 |
| tripropylene glycol | 192.3 | 1.110 | supercools | 265.1 |
| 1,3-propanediol | 76.1 | 1.060 | -28 | 210 |
| 1,3-butanediol | 92.1 | 1.005 | - | 207.5 |
| 1,4-butanediol | 92.1 | 1.017 | 20.1 | 235 |
| neopentyl glycol | 104.2 | - | 130 | 206 |
| 1,6-hexanediol | 118.2 | 1.017 | 43 | 250 |
| 1,4-cyclohexanedimethanol | - | - | - | - |
| HQEE | - | - | - | - |
| ethanolamine | 61.1 | 1.018 | 10.3 | 170 |
| diethanolamine | 105.1 | 1.097 | 28 | 271 |
| methyldiethanolamine | 119.1 | 1.043 | -21 | 242 |
| phenyldiethanolamine | 181.2 | - | 58 | 228 |
| ***hydroxyl compounds – trifunctional molecules*** | | | | |
| | MW | s.g. | f.p. °C | b.p. °C |
| glycerol | 92.1 | 1.261 | 18.0 | 290 |
| trimethylolpropane | - | - | - | - |
| 1,2,6-hexanetriol | - | - | - | - |
| triethanolamine | 149.2 | 1.124 | 21 | - |
| ***hydroxyl compounds – tetrafunctional molecules*** | | | | |
| | MW | s.g. | m.p. °C | b.p. °C |
| pentaerythritol | 136.2 | - | 260.5 | - |
| N,N,N',N'-tetrakis(2-hydroxypropyl)ethylenediamine | - | - | - | - |
| ***amine compounds – difunctional molecules*** | | | | |
| | MW | s.g. | m.p. °C | b.p. °C |
| diethyltoluenediamine | 178.3 | 1.022 | - | 308 |
| dimethylthiotoluenediamine | 214.0 | 1.208 | - | - |

## *Catalysts*

Polyurethane catalysts can be classified into two broad categories, amine compounds and organometallic complexes. They can be further classified as to their specificity, balance, and relative power or efficiency. Traditional amine catalysts have been tertiary amines such as triethylenediamine (TEDA, also known as 1,4-diazabicyclo octane or DABCO, an Air Products's trade mark), dimethylcyclohexylamine (DMCHA), and dimethylethanolamine (DMEA).

Tertiary amine catalysts are selected based on whether they drive the urethane (polyol+isocyanate, or gel) reaction, the urea (water+isocyanate, or blow) reaction, or the isocyanate trimerization reaction (e.g. using potassium acetate, to form isocyanurate ring structure). Since most tertiary amine catalysts will drive all three reactions to some extent, they are also selected based on how much they favour one reaction over another.

For example, tetramethylbutanediamine (TMBDA) preferentially drives the gel reaction over the blow reaction. On the other hand, both pentamethyldipropylenetriamine and N-(3-dimethylaminopropyl)-N,N-diisopropanolamine balance the blow and gel reactions, although the former is more potent than the later on a weight basis. 1,3,5-(tris(3-dimethylamino)propyl)-hexahydro-s-triazine is a trimerization catalyst that also strongly drives the blow reaction.

Molecular structure gives some clue to the strength and selectivity of the catalyst. Blow catalysts generally have an ether linkage two carbons away from a tertiary nitrogen. Examples include bis-(2-dimethylaminoethyl)ether (also known as A-99, formerly a Union Carbide product), and N-ethylmorpholine. Strong gel catalysts contain alkyl-substituted nitrogens, such as triethylamine (TEA), 1,8-diazabicyclo undecene-7 (DBU), and pentamethyldiethylenetriamine (PMDETA). Weaker gel catalysts contain ring-substituted nitrogens, such as benzyldimethylamine (BDMA). Trimerization catalysts contain the triazine structure, or are quaternary ammonium salts. Two trends have emerged since the late 1980s.

The requirement to fill large, complex tooling with increasing production rates has led to the use of blocked catalysts to delay front end reactivity while maintaining back end cure. In the United States, acid- and quaternary ammonium salt-blocked TEDA and bis-(2-dimethylaminoethyl) ether are common blocked catalysts used in molded flexible foam and microcellular integral skin foam applications.

Increasing aesthetic and environmental awareness has led to the use of non-fugitive catalysts for vehicle interior and furnishing applications in order to reduce odor, fogging, and the staining of vinyl coverings. Catalysts that contain a hydroxyl group or an active amino hydrogen, such as N,N,N'-trimethyl-N'-hydroxyethyl-bis(aminoethyl)ether and N'-(3-(dimethylamino)propyl)-N, N-dimethyl-1, 3-propanediamine that react into the polymer matrix can replace traditional catalysts in these applications.

Organometallic compounds based on mercury, lead, tin (dibutyltin dilaurate), bismuth (bismuth octanoate), and zinc are used as polyurethane catalysts. Mercury carboxylates, such as phenylmercuric neodeconate, are particularly effective catalysts for polyurethane elastomer, coating and sealant applications, since they are very highly selective towards the polyol+isocyanate reaction. Mercury catalysts can be used at low levels to give systems a long pot life while still giving excellent back-end cure. Lead catalysts are used in highly reactive rigid spray foam insulation applications, since they maintain their potency in low temperature and high humidity conditions. Due to their toxicity and the necessity to dispose of mercury and lead catalysts and catalysed material as hazardous waste in the United States, formulators have been searching for suitable replacements. Since the 1990s, bismuth and zinc carboxylates have been used as alternatives but have shortcomings of their own. In elastomer applications, long pot life systems do not build green strength as fast as mercury catalysed systems. In spray foam applications, bismuth and zinc do not drive the front end fast enough in cold weather conditions and must be otherwise augmented to replace lead. Alkyl tin carboxylates, oxides and mercaptides oxides are used in all types of polyurethane applications. For example, dibutyltin dilaurate is a standard catalyst for polyurethane adhesives and sealants, dioctyltin mercaptide is used in microcellular elastomer applications, and dibutyltin oxide is used in polyurethane paint and coating applications. Tin mercaptides are used in formulations that contain water, as tin carboxylates are susceptible to degradation from hydrolysis.

### Surfactants

Surfactants are used to modify the characteristics of both foam and non-foam polyurethane polymers. They take the form of polydimethylsiloxane-polyoxyalkylene block copolymers, silicone oils, nonylphenol ethoxylates, and other organic compounds. In foams, they are used to emulsify the liquid components, regulate cell size, and stabilize the cell structure to prevent collapse and sub-surface voids. In non-foam applications they are used as air release and anti-foaming agents, as wetting agents, and are used to eliminate surface defects such as pin holes, orange peel, and sink marks.

## Production

The main polyurethane producing reaction is between a diisocyanate (aromatic and aliphatic types are available) and a polyol, typically a polypropylene glycol or polyester polyol, in the presence of catalysts and materials for controlling the cell structure, (surfactants) in the case of foams. Polyurethane can be made in a variety of densities and hardnesses by varying the type of monomer(s) used and adding other substances to modify their characteristics, notably density, or enhance their performance. Other additives can be used to improve the fire performance, stability in difficult chemical environments and other properties of the polyurethane products.

## Health and Safety

Fully reacted polyurethane polymer is chemically inert. No exposure limits have been established by OSHA (Occupational Safety and Health Administration) or ACGIH (American Conference of Governmental Industrial Hygienists). It is not regulated by OSHA for carcinogenicity. Polyurethane polymer is a combustible solid and can be ignited if exposed to an open flame. Decomposition from fire can produce mainly carbon monoxide, and trace nitrogen oxides and hydrogen cyanide. Firefighters should wear self-contained breathing apparatus in enclosed areas.

Liquid resin blends and isocyanates may contain hazardous or regulated components. They should be handled in accordance with manufacturer recommendations found on product labels, and in MSDS (Material Safety Data Sheet) and product technical literature. Isocyanates are known skin and respiratory sensitizers, and proper engineering controls should be in place to prevent exposure to isocyanate liquid and vapour. Proper hygiene controls and personal protective equipment (PPE), such as gloves, respirators, and protective clothing and eye wear should be used.

In the United States, additional health and safety information can be found through organizations such as the Polyurethane Manufacturers Association (PMA) and the Centre for the Polyurethanes

Industry (CPI), as well as from polyurethane system and raw material manufacturers. In Europe, health and safety information is available from ISOPA, the European Diisocyanate and Polyol Producers Association. Regulatory information can be found in the Code of Federal Regulations Title 21 (Food and Drugs) and Title 40 (Protection of the Environment).

### Manufacturing

The methods of manufacturing polyurethane finished goods range from small, hand pour piece-part operations to large, high-volume bunstock and boardstock production lines. Regardless of the end-product, the manufacturing principle is the same: to meter the liquid isocyanate and resin blend at a specified stoichiometric ratio, mix them together until a homogeneous blend is obtained, dispense the reacting liquid into a mold or on to a surface, wait until it cures, then demold the finished part.

### Dispensing Equipment

Although the capital outlay can be high, it is desirable to use a meter-mix or dispense unit for even low-volume production operations that require a steady output of finished parts. Dispense equipment consists of material holding (day) tanks, metering pumps, a mix head, and a control unit. Often, a conditioning or heater-chiller unit is added to control material temperature in order to improve mix efficiency, cure rate, and to reduce process variability. Choice of dispense equipment components depends on shot size, throughput, material characteristics such as viscosity and filler content, and process control. Material day tanks may be single to hundreds of gallons in size, and may be supplied directly from drums, IBCs (intermediate bulk containers, such as totes), or bulk storage tanks. They may incorporate level sensors, conditioning jackets, and mixers. Pumps can be sized to meter in single grams per second up to hundreds of pounds per minute. They can be rotary, gear, or piston pumps, or can be specially hardened lance pumps to meter liquids containing highly abrasive fillers such as wollastonite, chopped or hammer milled glass fibres.

### Tooling

Distinct from pour-in-place, bun and boardstock, and coating applications, the production of piece parts requires some type of tooling to contain and form the reacting liquid. The choice of mold-making material is dependent on the expected number of uses to end-of-life (EOL), molding pressure, flexibility, and heat transfer characteristics. RTV silicone is used for tooling that has an EOL in the thousands of parts. It is typically used for molding rigid foam parts, where the ability to stretch and peel the mold around undercuts is needed. The heat transfer characteristic of RTV silicone tooling is poor. High-performance, flexible polyurethane elastomers are also used in this way.

Epoxy, metal-filled epoxy, and metal-coated epoxy is used for tooling that has an EOL in the tens-of-thousands of parts. It is typically used for molding flexible foam cushions and seating, integral skin and microcellular foam padding, and shallow-draft RIM bezels and fascia.

The heat transfer characteristic of epoxy tooling is fair; the heat transfer characteristic of metal-filled and metal-coated epoxy is good. Copper tubing can be incorporated into the body of the tool, allowing hot water to circulate and heat the mold surface. Aluminum is used for tooling that has an EOL in the hundreds-of-thousands of parts. It is typically used for molding microcellular foam gasketing and cast elastomer parts, and is milled or extruded into shape. Mirror-finish stainless steel is used for tooling that imparts a glossy appearance to the finished part. The heat transfer characteristic of metal tooling is excellent. Finally, molded or milled polypropylene is used to create low-volume tooling for molded gasket applications. Instead of many expensive metal molds, low-cost plastic tooling can be formed from a single metal master, which also allows greater design flexibility. The heat transfer characteristic of polypropylene tooling is poor, which must be taken into consideration during the formulation process.

## *Applications*

In 2007, the global consumption of polyurethane raw materials was above 12 million metric tons, the average annual growth rate is about 5%.

## *Testing—Effects of Visible Light*

Polyurethanes, especially those made using aromatic isocyanates, contain chromophores which interact with light. This is of particular interest in the area of polyurethane coatings, where light stability is a critical factor and is the main reason that aliphatic isocyanates are used in making polyurethane coatings. When PU foam, which is made using aromatic isocyanates, is exposed to visible light it discolors, turning from off-white to yellow to reddish brown. It has been generally accepted that apart from yellowing, visible light has little effect on foam properties. This is especially the case if the yellowing happens on the outer portions of a large foam, as the deterioration of properties in the outer portion has little effect on the overall bulk properties of the foam itself.

It has been reported that exposure to visible light can affect the variability of some physical property test results. Increasing exposure time and/or light intensity during the storage of foam samples under ambient laboratory conditions increased the amount of permanent set induced in some compression set tests (the samples did not fully return to their original size and/or shape). Variability resulted from uncontrolled light exposure of cut samples prior to being compressed. Other foam properties were not substantively affected. It was recommended that specimen preparation and testing be done rapidly to minimize variation in results or if specimens are prepared but not tested for a week or more, that the samples should be protected from light exposure.

Higher-energy UV radiation promotes chemical reactions in foam, some of which are detrimental to the foam structure.

## Polyurea

Polyurea is a type of elastomer that is derived from the reaction product of an isocyanate component and a synthetic resin blend component through step-growth polymerization. The isocyanate can be aromatic or aliphatic in nature. It can be monomer, polymer, or any variant reaction of isocyanates, quasi-prepolymer or a prepolymer. The prepolymer, or quasi-prepolymer, can be made of an amine-terminated polymer resin, or a hydroxyl-terminated polymer resin. The resin blend may be made up of amine-terminated polymer resins, and/or amine-terminated chain extenders. The amine-terminated polymer resins will not have any intentional hydroxyl moieties. Any hydroxyls are the result of incomplete conversion to the amine-terminated polymer resins. The resin blend may also contain additives, or non-primary components. These additives may contain hydroxyls, such as pre-dispersed pigments in a polyol carrier. Normally, the resin blend will not contain a catalyst(s).

### *Polymer Structure*

General reaction for forming a polyurea chain, illustrating the two monomer reactants and highlighting the urea linkage in the product.

The latter term refers to the substance urea, found in urine, rather than urine itself. Urea or carbamide is an organic compound with the chemical formula $(NH_2)_2CO$. The molecule has two amine groups ($-NH_2$) joined by a carbonyl functional group (C=O). In a polyurea, alternating monomer units of isocyanates and amines react with each other to form urea linkages. Ureas can also be formed from the reaction of isocyanates and water which forms a carbamic acid intermediate. This acid quickly decomposes by splitting off carbon dioxide and leaving behind an amine. This amine then reacts with another isocyanate group to form the polyurea linkage. This two step reaction is used in what is commonly but improperly called polyurethane foams. The carbon dioxide that is liberated in this reaction is the primary blowing (foaming) agent especially in many polyurethane foams which more precisely should be called polyurethane/urea foams.

### Uses

Polyurea and polyurethane are copolymers used in the manufacture of spandex, which was invented in 1959.

Development of two-component polyurethane and polyurea spray elastomers took place in the 1990s. Their fast reactivity and relative insensitivity to moisture make them useful coatings for large surface area projects, such as secondary containment, manhole and tunnel coatings, tank liners, and truck bed liners. Excellent adhesion to concrete and steel is obtained with the proper primer and surface treatment. They can also be used for spray molding and armor. Some polyureas reach strengths of 6000psi (40MPa) tensile and over 500% elongation making it a tough coating. The quick cure time allows many coats to be built up quickly.

## Diol

A diol or glycol is a chemical compound containing two hydroxyl groups (—OH groups).

A geminal diol has two hydroxyl groups bonded to the same atom. Examples include methanediol $H_2C(OH)_2$ and 1,1,1,3,3,3-hexafluoropropane-2,2-diol $(F3C)_2C(OH)_2$, the hydrated form of hexafluoroacetone.

A vicinal diol is a diol with two hydroxyl groups in vicinal positions, that is, attached to adjacent atoms. Examples include 1,2-ethanediol or ethylene glycol HO—$(CH_2)_2$—OH, a common ingredient of antifreeze products; and propane-1,2-diol or alpha propylene glycol, HO—$CH_2$—CH(OH)—$CH_3$ used in the food and medicine industry as well as a relatively non-poisonous antifreeze product.

Examples of diols in which the hydroxyl functional groups are more widely separated include 1,4-butanediol HO—$(CH_2)_4$—OH and bisphenol A, and propylene-1,3-diol or beta propylene glycol, HO-$CH_2$-$CH_2$-$CH_2$-OH.

### Classification

Diols can be:

* Linear or branched;
* Aliphatic or aromatic (bisphenol A).

### Synthesis

Because diols are a common functional group arrangement, numerous methods of preparation have been developed.

* Vicinal diols can be produced from the oxidation of alkenes, usually with dilute acidic potassium permanganate, also known as potassium manganate(VII). Using alkaline potassium manganate(VII) produces a colour change from clear deep purple to clear green; acidic potassium manganate(VII) turns clear colourless.
* Osmium tetroxide can similarly be used to oxidize alkenes to vicinal diols.
* Hydrogen peroxide reacts with an alkene to the epoxide and then by saponification to the diol for example in the synthesis of trans-cyclohexanediol batch or by microreactor:

### Cyclohexanediol Synthesis

* A chemical reaction called Sharpless asymmetric dihydroxylation can be used to produce chiral diols from alkenes using an osmate reagent and a chiral catalyst.
* Another method is the Woodward cis-hydroxylation (cis diol) and the related Prévost reaction (anti diol), depicted below, which both use iodine and the silver salt of a carboxylic acid.

### Prevost Reaction

* In the Prins reaction 1,3-diols can be formed in a reaction between an alkene and formaldehyde.
* Geminal diols can be formed by the hydration of ketones.

### Reactions—General Diols

Diols react as alcohols, by esterification and ether formation.

Diols such as ethylene glycol are used as co-monomers in polymerization reactions forming polymers including some polyesters and polyurethanes. A different monomer with two identical functional groups, such as a dioyl dichloride or dioic acid is required to continue the process of polymerization through repeated esterification processes.

A diol can be converted to cyclic ether by using an acid catalyst, this is diol cyclization. Firstly, it involves protonation of the hydroxyl group. Then, followed by intramolecular nucleophilic substitution, the second hydroxyl group attacks the electron deficient carbon. Provided that there are enough carbon atoms that the angle strain is not too much, a cyclic ether can be formed.

### Vicinal Diols

In glycol cleavage, the C-C bond in a vicinal diol is cleaved with formation of ketone or aldehyde functional groups.

### Geminal Diols

In general, organic geminal diols readily dehydrate to form a carbonyl group. For example, carbonic acid ($(HO)_2C=O$) is unstable and has a tendency to convert to carbon dioxide ($CO_2$) and water ($H_2O$). Nevertheless, in rare situations the chemical equilibrium is in favour of the geminal diol. For example, when formaldehyde ($H_2C=O$) is dissolved in water the geminal diol ($H_2C(OH)_2$), methanediol, is favoured. Other examples are the cyclic geminal diols decahydroxycyclopentane ($C_5(OH)_{10}$) and dodecahydroxycyclohexane ($C_6(OH)_{12}$), which are stable, whereas the corresponding oxocarbons ($C_5O_5$ and $C_6O_6$) do not seem to be.

## Detergent

A detergent is a surfactant or a mixture of surfactants having "cleaning properties in dilute solutions." In common usage, "detergent" refers to alkylbenzenesulfonates, a family of compounds that are similar to soap but are less affected by hard water. In most household contexts, the term *detergent* by itself refers specifically to *laundry detergent* or *dish detergent*, as opposed to *hand soap* or other types of cleaning agents. Detergents are commonly available as powders or concentrated solutions. Detergents work because they are amphiphilic - partly hydrophilic (polar) and partly hydrophobic (non-polar). Their dual nature facilitates the mixture of hydrophobic compounds (like oil and grease) with water. Because air is not hydrophillic, detergents are also foaming agents to varying degrees. Completely non-polar solvents known as degreasers can also remove hydrophobic contaminants, but lacking polar elements may not dissolve in water.

### Chemical Classification of Detergents

Detergents are classified into three broad groupings, depending on the electrical charge of the surfactants.

### Anionic Detergents

Typical anionic detergents are alkylbenzenesulfonates. The alkylbenzene portion of these anions is lipophilic and the sulfonate is hydrophilic. Two varieties have been popularized, those with branched alkyl groups and those with linear alkyl groups. The former were largely phased out in economically advanced societies because they are poorly biodegradable. An estimated 6 billion kilograms of anionic detergents are produced annually for domestic markets. Bile acids, such as deoxycholic acid (DOC), are anionic detergents produced by the liver to aid in digestion and absorption of fats and oils.

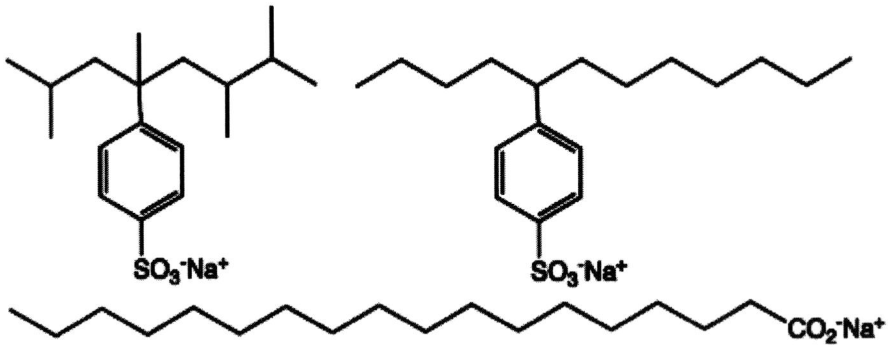


*Three kinds of anionic detergents: a branched sodium dodecylbenzenesulfonate, linear sodium dodecylbenzenesulfonate, and a soap.*

### Cationic Detergents

Cationic detergents are similar to the anionic ones, with a hydrophobic component, but instead of the anionic sulfonate group, the cationic surfactants have quaternary ammonium as the polar end. The ammonium centre is positively charged.

### *Non-ionic and Zwitterionic Detergents*

Non-ionic detergents are characterized by their uncharged, hydrophilic headgroups. Typical non-ionic detergents are based on polyoxyethylene or a glycoside. Common examples of the former include Tween, Triton, and the Brij series. These materials are also known as ethoxylates or PEGylates). Glycosides have a sugar as their uncharged hydrophilic headgroup. Examples include octyl-thioglucoside and maltosides. HEGA and MEGA series detergents are similar, possessing a sugar alcohol as headgroup.

Zwitterionic detergents possess a net zero charge arising from the presence of equal numbers of +1 and -1 charged chemical groups. Examples include CHAPS.

## Major Applications of Detergents

### *Laundry Detergents*

One of the largest applications of detergents is for cleaning clothing. The formulations are complex, reflecting the diverse demands of the application and the highly competitive consumer market. In general, laundry detergents contain water softeners, surfactants, bleach, enzymes, brighteners, fragrances, and many other agents. The formulation is strongly affected by the temperature of the cleaning water and varies from country to country.

### *Fuel Additives*

Both carburetors and fuel injector components of Otto engines benefit from detergents in the fuels to prevent fouling. Concentrations are about 300 ppm. Typical detergents are long chain amines and amides such as polyisobuteneamine and polyisobuteneamide/ succinimide.

### *Soapless Soap*

Soapless soap refers to a soap free liquid cleanser with a slightly acidic pH. Soapless soaps are used in an array of products.

## Sulfonate

A sulfonate is a salt or ester of a sulfonic acid. It contains the functional group $R$-$SO_2O$-.

### Sulfonate Salts

Anions with the general formula $RSO_2O^-$ are called sulfonates. They are the conjugate bases of sulfonic acids with formula $RSO_2OH$. As sulfonic acids tend to be strong acids, the corresponding sulfonates are weak bases. Due to the stability of sulfonate anions, the cations of sulfonate salts such as scandium triflate have application as Lewis acids.

A classic organic reaction for the preparation of sulfonates is that of alkyl halides with sulfites such as sodium sulfite, first described by Adolph Strecker in 1868 (Strecker sulfite alkylation). The general reaction is:

$$RX + M_2SO_3 \rightarrow RSO_3M + MX$$

Iodide is used as a catalyst.

### Sulfonic Esters

Esters with the general formula $R^1SO_2OR^2$ are as a category called *sulfonic esters*, with individual members of the category being named analogously to how ordinary carboxyl esters are named. For example, if the $R^2$ group is a methyl group and the $R^1$ group is a trifluoromethyl group, the resulting compound is methyl trifluoromethanesulfonate.

Sulfonic esters are used as reagents in organic synthesis, chiefly because the $RSO_2O$- group is a good leaving group in Sn1, Sn2, E1 and E2 reactions. Methyl triflate, for example, is a strong methylating reagent. They are commonly used to lend water solubility to protein crosslinkers such as N-hydroxysulfosuccinimide (Sulfo-NHS), BS3, Sulfo-SMCC, etc.

### Examples

- Mesylate (methanesulfonate), $CH_3SO_2O$-
- Triflate (trifluoromethanesulfonate), $CF_3SO_2O$-
- Tosylate (*p*-toluenesulfonate), $CH_3C_6H_4SO_2O$-
- Besylate (benzenesulfonate), $C_6H_5SO_2O$-

### Surfactant

Surfactants are compounds that lower the surface tension of a liquid, the interfacial tension between two liquids, or that between a liquid and a solid. Surfactants may act as detergents, wetting agents, emulsifiers, foaming agents, and dispersants.

### *Etymology*

The term *surfactant* is a blend of *surface active agents.*

In Index Medicus and the United States National Library of Medicine, *surfactant* is reserved for the meaning pulmonary surfactant. For the more general meaning, *surface active agent* is the heading.

### *Properties*

Surfactants are usually organic compounds that are amphiphilic, meaning they contain both hydrophobic groups (their *tails*) and hydrophilic groups (their *heads*). Therefore, a surfactant molecule contains both a water insoluble (or oil soluble) component and a water soluble component. Surfactant molecules will migrate to the water surface, where the insoluble hydrophobic group may extend out of the bulk water phase, either into the air or, if water is mixed with an oil, into the oil phase, while the water soluble head group remains in the water phase.

This alignment and aggregation of surfactant molecules at the surface acts to alter the surface properties of water at the water/air or water/oil interface. Because air is not hydrophillic, surfactants are also foaming agents to varying degrees. Completely non-polar solvents known as degreasers can also remove hydrophobic contaminants, but lacking polar elements may not dissolve in water.

Surfactants reduce the surface tension of water by adsorbing at the liquid-gas interface. They also reduce the interfacial tension their tails form a core that can encapsulate an oil droplet, and their (ionic/ polar) heads form an outer shell that maintains favourable contact with water. When surfactants assemble in oil, the aggregate is referred to as a reverse micelle. In a reverse micelle, the heads are in the core and the tails maintain favourable contact with oil. Surfactants are also often classified into four primary groups; anionic, cationic, non-ionic, and zwitterionic (dual charge).

Thermodynamics of the surfactant systems are of great importance, theoretically and practically. This is because surfactant systems represent systems between ordered and disordered states of matter. Surfactant solutions may contain an ordered phase (micelles) and a disordered phase (free surfactant molecules and/or ions in the solution).

The spectroscopy and analytical chemistry of individual surfactant compounds is given in multiple references, including a multi-volume analysis reference textbook

### *Applications and Sources*

Surfactants play an important role as cleaning, wetting, dispersing, emulsifying, foaming and anti-foaming agents in many practical applications and products, including:

- Detergents
- Fabric softeners
- Emulsions
- Paints
- Adhesives
- Inks
- Anti-fogs
- Ski waxes, snowboard wax
- Deinking of recycled papers, in flotation, washing and enzymatic processes
- Laxatives
- Agrochemical formulations
  - o Herbicides (some)
  - o Insecticides
- Quantum dot coatings
- Biocides (sanitizers)
- Cosmetics:
  - o Shampoos
  - o Hair conditioners (after shampoo)
  - o Toothpastes
- Spermicides (nonoxynol-9)
- Firefighting
- Pipelines, liquid drag reducing agent
- Alkali Surfactant Polymers (used to mobilize oil in oil wells)
- Ferrofluids
- Leak Detectors.

### *Detergents in Biochemistry and Biotechnology*

In solution: detergents help solubilize molecules by dissociating aggregates and unfolding proteins. These include SDS, CTAB. Detergents are key reagents to extract protein by lysis of the cells and tissues: they disorganize the membrane's lipidic bilayer (SDS, Triton X-100, X-114, CHAPS, DOC, NP-40), and solubilize proteins. Milder detergents such as (OctylThioGlucosides) are used to solubilize sensible proteins (enzymes, receptors). Non-solubilized material is harvested by centrifugation or other means. For electrophoresis for example, proteins are classically treated with SDS to denature the native tertiary and quaternary structures, allowing the separation of proteins according to their molecular weight.

Detergents have also been used to decellularise organs. This process maintains a matrix of proteins that preserves the structure of the organ and often the microvascular network. The process has been successfully used to prepare organs such as the liver and heart for transplant in rats. Pulmonary surfactants are also naturally secreted by type II cells of the lung alveoli in mammals.

### *According to the Composition of their Tail*

The tail of surfactants can be:

- A hydrocarbon chain: aromatic hydrocarbons (arenes), alkanes (alkyl), alkenes, cycloalkanes, alkyne-based;
- An alkyl ether chain:
    - o Ethoxylated surfactants: polyethylene oxides are inserted to increase the hydrophilic character of a surfactant;
    - o Propoxylated surfactants: polypropylene oxides are inserted to increase the lipophilic character of a surfactant;
- A fluorocarbon chain: fluorosurfactants;
- A siloxane chain: siloxane surfactants

A surfactant can have one or two tails, these are called double-chained.

### *According to the Composition of their Head*

A surfactant can be classified by the presence of formally charged groups in its head. A non-ionic surfactant has no charge groups in its head. The head of an ionic surfactant carries a net charge. If the charge is negative, the surfactant is more specifically called anionic; if the charge is positive, it is called cationic. If a surfactant contains a head with two oppositely charged groups, it is termed zwitterionic.

Some commonly encountered surfactants of each type include:

- Ionic

- Anionic: based on permanent anions (sulfate, sulfonate, phosphate) or pH-dependent anions (carboxylate):
  - Sulfates:
  - Alkyl sulfates: ammonium lauryl sulfate, sodium lauryl sulfate (SDS, sodium dodecyl sulfate, another name for the compound);
  - Alkyl ether sulfates: sodium laureth sulfate, also known as sodium lauryl ether sulfate (SLES), sodium myreth sulfate;
  - Sulfonates:
  - Docusates: dioctyl sodium sulfosuccinate;
  - Sulfonate fluorosurfactants: perfluorooctanesulfonate (PFOS), perfluorobutanesulfonate;
  - Alkyl benzene sulfonates;
  - Phosphates:
  - Alkyl aryl ether phosphate
  - Alkyl ether phosphate
  - Carboxylates:
  - Alkyl carboxylates: Fatty acid salts (soaps): sodium stearate;
  - Sodium lauroyl sarcosinate;
  - Carboxylate fluorosurfactants: perfluorononanoate, perfluorooctanoate (PFOA or PFO)
- Cationic: based on:
  - pH-dependent primary, secondary or tertiary amines: primary amines become positively charged at pH < 10, secondary amines become charged at pH < 4:
  - Octenidine dihydrochloride;
  - Permanently charged quaternary ammonium cation:
  - Alkyltrimethylammonium salts: cetyl trimethylammonium bromide (CTAB) a.k.a. hexadecyl trimethyl ammonium bromide, cetyl trimethylammonium chloride (CTAC);
  - Cetylpyridinium chloride (CPC);
  - Polyethoxylated tallow amine (POEA);
  - Benzalkonium chloride (BAC);

  * Benzethonium chloride (BZT);
  * 5-Bromo-5-nitro-1,3-dioxane;
  * Dimethyldioctadecylammonium chloride
  * Dioctadecyldimethylammonium bromide (DODAB)
- o Zwitterionic (amphoteric): based on primary, secondary or tertiary amines or quaternary ammonium cation with:
  * Sulfonates:
  * CHAPS (3-[(3-Cholamidopropyl)dimethylammonio]-1-propanesulfonate);
  * Sultaines: cocamidopropyl hydroxysultaine;
  * Carboxylates:
  * Amino acids
  * Imino acids
  * Betaines: cocamidopropyl betaine;
  * Phosphates: lecithin

• Nonionic
- o Fatty alcohols:
  * Cetyl alcohol,
  * Stearyl alcohol,
  * Cetostearyl alcohol (consisting predominantly of cetyl and stearyl alcohols),
  * Oleyl alcohol;
- o Polyoxyethylene glycol alkyl ethers (Brij): $CH_3–(CH_2)_{10–16}–(O-C_2H_4)_{1–25}–OH$:
  * Octaethylene glycol monododecyl ether,
  * Pentaethylene glycol monododecyl ether;
- o Polyoxypropylene glycol alkyl ethers: $CH_3–(CH_2)_{10–16}–(O-C_3H_6)_{1–25}–OH$;
- o Glucoside alkyl ethers: $CH_3–(CH_2)_{10–16}$–(O-Glucoside)$_{1–3}$–OH:
  * Decyl glucoside,
  * Lauryl glucoside,
  * Octyl glucoside;
- o Polyoxyethylene glycol octylphenol ethers: $C_8H_{17}–(C_6H_4)–(O-C_2H_4)_{1–25}–OH$:

        * Triton X-100;
    - o Polyoxyethylene glycol alkylphenol ethers: $C_9H_{19}$–($C_6H_4$)–(O-$C_2H_4$)$_{1-25}$–OH:
        * Nonoxynol-9;
    - o Glycerol alkyl esters:
        * Glyceryl laurate
    - o Polyoxyethylene glycol sorbitan alkyl esters: Polysorbates;
    - o Sorbitan alkyl esters: Spans;
    - o Cocamide MEA, cocamide DEA;
    - o Dodecyldimethylamine oxide;
    - o Block copolymers of polyethylene glycol and polypropylene glycol: Poloxamers

### *According to the Composition of their Counter-ion*

In the case of ionic surfactants, the counter-ion can be:

- Monoatomic/Inorganic:
    - o Cations: metals : alkali metal, alkaline earth metal, transition metal;
    - o Anions: halides: chloride ($Cl^-$), bromide ($Br^-$), iodide ($I^-$);
- Polyatomic/Organic:
    - o Cations: ammonium, pyridinium, triethanolamine (TEA)
    - o Anions: tosyls, trifluoromethanesulfonates, methylsulfate.

### *Current Market*

The annual global production of surfactants was 13 million metric tons in 2008 and the annual turnover reached US$24.33 billion in 2009, nearly 2% up from the previous year. The market is expected to experience quite healthy growth by 2.8% annually to 2012 and by 3.5 - 4% thereafter.

### *Health and Environmental Controversy*

Some surfactants are known to be toxic to animals, ecosystems and humans, and can increase the diffusion of other environmental contaminants. Despite this, they are routinely deposited in numerous ways on land and into water systems, whether as part of an intended process or as industrial and household waste. Some surfactants have proposed or voluntary restrictions on their use. For example, PFOS is a persistent organic pollutant as judged by the Stockholm Convention.

Additionally, PFOA has been subject to a voluntary agreement by the U.S. Environmental Protection Agencyý and eight chemical companies to reduce and eliminate emissions of the chemical and its precursors.

The two major surfactants used in the year 2000 were linear alkylbenzene sulphonates (LAS) and the alkyl phenol ethoxylates (APE). They break down in the aerobic conditions found in sewage treatment plants and in soil.

Ordinary dishwashing detergent, for example, will promote water penetration in soil, but the effect would only last a few days (many standard laundry detergent powders contain levels of chemicals such as alkali and chelating agents, which can be damaging to plants and should not be applied to soils). Commercial soil wetting agents will continue to work for a considerable period, but they will eventually be degraded by soil micro-organisms. Some can, however, interfere with the life-cycles of some aquatic organisms, so care should be taken to prevent run-off of these products into streams, and excess product should not be washed down.

Anionic surfactants can be found in soils as the result of sludge application, wastewater irrigation, and remediation processes. Relatively high concentrations of surfactants together with multimetals can represent an environmental risk. At low concentrations, surfactant application is unlikely to have a significant effect on trace metal mobility.

### Biosurfactants

Biosurfactants are surface-active substances synthesised by living cells; they are generally non-toxic and biodegradable. Interest in microbial surfactants has been steadily increasing in recent years due to their diversity, environmentally friendly nature, possibility of large-scale production, selectivity, performance under extreme conditions and potential applications in environmental protection.

Biosurfactants enhance the emulsification of hydrocarbons, have the potential to solubilise hydrocarbon contaminants and increase their availability for microbial degradation. The use of chemicals for the treatment of a hydrocarbon polluted site may contaminate the environment with their by-products, whereas biological treatment may efficiently destroy pollutants, while being biodegradable themselves. Hence, biosurfactant producing microorganisms may play an important role in the accelerated bioremediation of hydrocarbon contaminated sites. These compounds can also be used in enhanced oil recovery and may be considered for other potential applications in environmental protection. Other applications include herbicides and pesticides formulations, detergents, health care and cosmetics, pulp and paper, coal, textiles, ceramic processing and food industries, uranium ore-processing and mechanical dewatering of peat.

Several microorganisms are known to synthesise surface-active agents, most of them are bacteria and yeasts. When grown on hydrocarbon substrate as the carbon source, these microorganisms synthesise a wide range of chemicals with surface activity, such as glycolipid, phospholipid and others. These chemicals are apparently synthesised to emulsify the hydrocarbon substrate and facilitate its transport into the cells. In some bacterial species such as *Pseudomonas aeruginosa*, biosurfactants are also involved in a group motility behaviour called swarming motility.

### *Biosurfactants and Deepwater Horizon*

The use of biosurfactants as a way to remove petroleum from contaminated sites has been questioned, and criticized as irresponsible and environmentally unsafe. Biosurfactants were not used by BP after the Deepwater Horizon offshore drilling rig went down on April 20, 2010, on the resulting Deepwater Horizon oil spill. However, unprecedented amounts of Corexit, a surfactant solution produced by Nalco Holding Company (whose active ingredient is Tween-80), were sprayed directly into the ocean at the leak and on the sea-water's surface, the theory being that the surfactants would isolate individual molecules of oil making it easier for petroleum consuming microbes to digest the oil. However some scientists say that rather than helping the situation the surfactants have only managed to disperse and sink the oil below the surface and out of sight. Naturally occurring petroleum consuming microbes have evolved on the bottom of the ocean where they have adapted to live in areas where oil seeps naturally from the ocean floor.

# 5

# Chlorobenzene

Chlorobenzene is an aromatic organic compound with the chemical formula $C_6H_5Cl$. This colorless, flammable liquid is a common solvent and a widely used intermediate in the manufacture of other chemicals.

### *Uses*

Chlorobenzene once was used in the manufacture of certain pesticides, most notably DDT by reaction with chloral (trichloroacetaldehyde), but this application has declined with the diminished use of DDT. At one time, chlorobenzene was the main precursor for the manufacture of phenol:

$$C_6H_5Cl + NaOH \rightarrow C_6H_5OH + NaCl$$

The major use of chlorobenzene is as an intermediate in the production of commodities such as herbicides, dyestuffs, and rubber. Chlorobenzene is also used as a high-boiling solvent in many industrial applications as well as in the laboratory. Chlorobenzene is nitrated on a large scale to give a mixture of 2- and 4-nitrochlorobenzenes, which can be separated by fractional crystallization followed by distillation. 2-Nitrochlorobenzene (CAS#88-73-3) is converted to related 2-nitrophenol, 2-nitroanisole, bis(2-nitrophenyl)disulfide, and 2-nitropaniline by nucleophilic displacement of the chloride with sodium hydroxide, sodium methoxide, sodium disulfide and ammonia. The conversion of the 4-nitrochlorobenzene (CAS#100-00-5) are similar.

### *Synthesis*

Chlorobenzene was first described in 1851. Presently it is manufactured by chlorination of benzene in the presence of a catalytic amount of Lewis acid such as ferric chloride and anhydrous aluminium chloride:

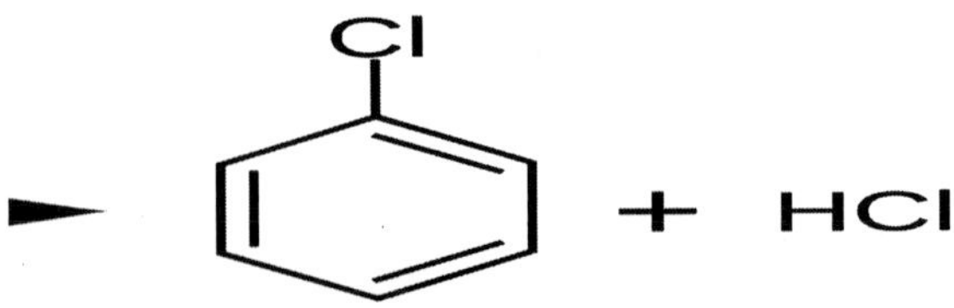

The catalyst enhances the electrophilicity of the chlorine. Because chlorine is electronegative, $C_6H_5Cl$ exhibits decreased susceptibility to attack by other electrophiles. For this reason, the chlorination process produces only small amounts of dichloro- and trichlorobenzenes.

### *Safety*

Chlorobenzene exhibits "low to moderate" toxicity as indicated by its $LD_{50}$ of 2.9 g/kg.

### *Toxicology*

Chlorobenzene is not a naturally occurring compound, but because of its manufacture, can persist in soil for several months, in air for about 3.5 days, and in water for less than one day. Humans may be exposed to this agent via breathing contaminated air (primarily via occupational exposure), eating contaminated food or water, or by coming into contact with contaminated soil (typically near hazardous waste sites). However, because it has only been found at 97 out of 1,177 NPL hazardous waste sites, it is not considered a widespread environmental contaminant.

Upon entering the body, typically via contaminated air, chlorobenzene is excreted both via the lungs and the urinary system.

## Toluene

Toluene, formerly known as toluol, is a clear, water-insoluble liquid with the typical smell of paint thinners. It is a mono-substituted benzene derivative, i.e., one in which a single hydrogen atom from the benzene molecule has been replaced by a univalent group, in this case $CH_3$.

It is an aromatic hydrocarbon that is widely used as an industrial feedstock and as a solvent. Like other solvents, toluene is sometimes also used as an inhalant drug for its intoxicating properties; however, inhaling toluene has potential to cause severe neurological harm. Toluene is an important organic solvent, but is also capable of dissolving a number of notable inorganic chemicals such as sulfur.

## *History*

The name *toluene* was derived from the older name *toluol*, which refers to tolu balsam, an aromatic extract from the tropical Colombian tree *Myroxylon balsamum*, from which it was first isolated. It was originally named by Jöns Jakob Berzelius.

## *Chemical Properties*

Toluene reacts as a normal aromatic hydrocarbon towards electrophilic aromatic substitution. The methyl group makes it around 25 times more reactive than benzene in such reactions. It undergoes smooth sulfonation to give *p*-toluenesulfonic acid, and chlorination by $Cl_2$ in the presence of $FeCl_3$ to give ortho and para isomers of chlorotoluene. It undergoes nitration to give ortho and para nitrotoluene isomers, but if heated it can give dinitrotoluene and ultimately the explosive trinitrotoluene (TNT).

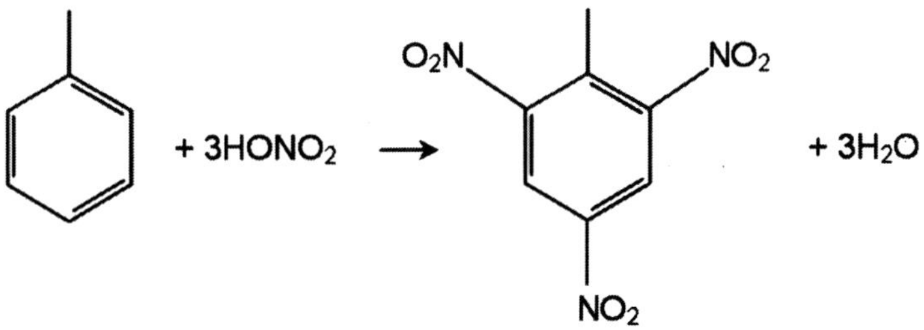


With other reagents the methyl side chain in toluene may react, undergoing oxidation. Reaction with basify potassium permanganate and diluted acid (e.g., sulfuric acid) or potassium permanganate with concentrated sulfuric acid, leads to benzoic acid, whereas reaction with chromyl chloride leads to benzaldehyde (Étard reaction). Halogenation can be performed under free radical conditions. For example, *N*-bromosuccinimide (NBS) heated with toluene in the presence of AIBN leads to benzyl bromide. Toluene can also be treated with elemental bromine in the presence of UV light (direct sunlight) to yield benzyl bromide. Toluene may also be brominated by treating it with HBr and $H_2O_2$ in the presence of light.

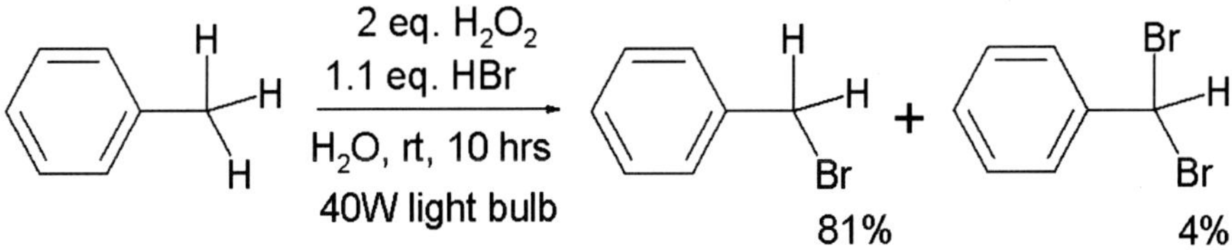

Catalytic hydrogenation of toluene to methylcyclohexane requires a high pressure of hydrogen to go to completion, because of the stability of the aromatic system. $pK_a$ is approximately 45.

## Production

Toluene occurs naturally at low levels in crude oil and is usually produced in the processes of making gasoline via a catalytic reformer, in an ethylene cracker or making coke from coal. Final separation (either via distillation or solvent extraction) takes place in a BTX plant.

## Preparation

***From benzene (Friedel–Crafts reaction):*** Benzene reacts with methyl chloride in presence of anhydrous aluminium chloride to form toluene. The formation follows an electrophilic substitution reaction mechanism:

$$CH_3Cl + AlCl_3 \rightarrow CH_3^+ + AlCl_4^-$$

$$C_6H_5H + CH_3^+ + AlCl_4^- \rightarrow C_6H_5CH_3 + HCl + AlCl_3$$

The following catalysts can be used in place of $AlCl_3$:

$$AlCl_3 > SbCl_3 > SnCl_4 > BF_3 > ZnCl_2 > HgCl_2$$

Note that the reaction is not very useful as the mono-alkyl derivative formed readily undergoes further alkylation at a still-greater speed to produce polysubstituted products.

***From bromobenzene (Wurtz-Fittig reaction):*** The Wurtz-Fittig reaction is the reaction of an aryl halide and alkyl halide in presence of sodium metal to give substituted aromatic compounds.

When bromobenzene and methyl bromide react with sodium metal in dry ether solution, toluene is obtained.

$$C_6H_5Br + CH_3Br + 2Na \rightarrow C_6H_5CH_3 + 2NaBr$$

***From toluic acid (decarboxylation):*** When sodium salt of toluic acid (o-, m-, p-) is heated with soda lime, toluene is obtained.

$C_6H_4CH_3COONa$ (sodium toluate) + NaOH $\rightarrow$ $C_6H_5CH_3$ (toluene) + $Na_2CO_3$

***From cresol:*** When cresol (o-, m-, p-) is distilled with zinc dust, toluene is obtained.

$C_6H_4CH_3OH$ (cresol) + Zn $\rightarrow$ $C_6H_5CH_3$ (toluene) + ZnO

***From toluenesulfonic acid:*** When toluenesulfonic acid is treated with superheated steam or boiled with HCl, toluene is obtained.

$CH_3C_6H_4SO_3H$ (toluenesulfonic acid) + HOH (steam) → $C_6H_5CH_3$ (toluene) + $H_2SO_4$ (sulfuric acid)

***From toluidine:*** Toluidine is first diazotized with sodium nitrite ($NaNO_2$) and HCl at low temperature. The diazonium compound thus obtained is heated with alkaline stannous chloride ($SnCl_2$). This reaction gives toluene.

***From Grignard reagent:*** When phenyl magnesium bromide ($C_6H_5$)MgBr is reacted with methyl bromide, toluene is obtained.

***From aromatic ketones:*** Using a Friedel–Crafts reaction, when an acid chloride is reacted with an aromatic hydrocarbon in the presence of anhydrous aluminium chloride ($AlCl_3$), a mixed aliphatic/aromatic ketone is obtained. The ketone is then reduced with amalgamated zinc and concentrated HCl.

$C_6H_6$ + $ClCOCH_3$ → $C_6H_5COCH_3$ (acetophenone) + HCl

$C_6H_5COCH_3$ + 4H → $C_6H_5CH_2CH_3$ (ethyl benzene).

This method is used to produce alkyl benzenes other than toluene.

## Uses

Toluene is a common solvent, able to dissolve paints, paint thinners, silicone sealants, many chemical reactants, rubber, printing ink, adhesives (glues), lacquers, leather tanners, and disinfectants. It can also be used as a fullerene indicator, and is a raw material for toluene diisocyanate (used in the manufacture of polyurethane foam) and TNT. In addition, it is used as a solvent to create a solution of carbon nanotubes. It is also used as a cement for fine polystyrene kits (by dissolving and then fusing surfaces) as it can be applied very precisely by brush and contains none of the bulk of an adhesive.

Industrial uses of toluene include dealkylation to benzene, and the disproportionation to a mixture of benzene and xylene in the BTX process. When oxidized it yields benzaldehyde and benzoic acid, two important intermediates in chemistry. It is also used as a carbon source for making Multi-Wall Carbon Nanotubes. Toluene can be used to break open red blood cells in order to extract hemoglobin in biochemistry experiments.

Toluene can be used as an octane booster in gasoline fuels used in internal combustion engines. Toluene at 86% by volume fuelled all the turbo Formula 1 teams in the 1980s, first pioneered by the Honda team. The remaining 14% was a "filler" of n-heptane, to reduce the octane to meet Formula 1 fuel restrictions. Toluene at 100% can be used as a fuel for both two-stroke and four-stroke engines; however, due to the density of the fuel and other factors, the fuel does not vaporize easily unless preheated to 70 degrees Celsius (Honda accomplished this in their Formula 1 cars by routing the fuel lines through the muffler system to heat the fuel). Toluene also poses similar problems as alcohol fuels, as it eats through standard rubber fuel lines and has no lubricating properties, as standard gasoline does, which can break down fuel pumps and cause upper cylinder bore wear.

In Australia, toluene has been found to have been illegally combined with petrol in fuel outlets for sale as standard vehicular fuel. Toluene attracts no fuel excise, while other fuels are taxed at over 40%, so fuel suppliers are able to profit from substituting the cheaper toluene for petrol. This substitution is likely to affect engine performance and result in additional wear and tear. The extent of toluene substitution has not been determined.

Toluene is another in a group of fuels that have recently been used as components for jet fuel surrogate blends. Toluene is used as a jet fuel surrogate for its content of aromatic compounds.

Toluene has also been used as a coolant for its good heat transfer capabilities in sodium cold traps used in nuclear reactor system loops.

Toluene had also been used in the process of removing the cocaine from coca leaves in the production of Coca-Cola syrup.

### Biology

Similar to many other solvents such as 1,1,1-trichloroethane and some alkylbenzenes, toluene has been shown to act as a non-competitive NMDA receptor antagonist and $GABA_A$ receptor positive allosteric modulator. It is abused as an inhalant likely on account of the euphoric and dissociative effects these actions produce. Additionally, toluene has been shown to display antidepressant-like effects in rodents in the forced swim test (FST) and the tail suspension test (TST).

### Toxicology and Metabolism

Toluene should not be inhaled due to its health effects. Low to moderate levels can cause tiredness, confusion, weakness, drunken-type actions, memory loss, nausea, loss of appetite, and hearing and colour vision loss. These symptoms usually disappear when exposure is stopped. Inhaling high levels of toluene in a short time may cause light-headedness, nausea, or sleepiness. It can also cause unconsciousness, and even death.

Toluene is, however, much less toxic than benzene, and has, as a consequence, largely replaced it as an aromatic solvent in chemical preparation.

## Toluene Diisocyanate

Toluene diisocyanate (TDI) is an aromatic diisocyanate. It is produced for reaction with polyols to form polyurethanes. It exists in two isomers, 2,4-TDI (CAS: 584-84-9) and 2,6-TDI (CAS: 91-08-7). 2,4-TDI is produced in the pure state, but TDI is often marketed as 80/20 and 65/35 mixtures of the 2,4 and 2,6 isomers respectively. It is a highly produced diisocyanate, accounting for 34.1% of the global isocyanate market in 2000, second only to MDI. All major producers of TDI are members of the International Isocyanate Institute, whose aim is the promotion of the safe handling of TDI in the workplace, community and environment.

### *Synthesis*

There are six steps to the synthesis of pure 2,4-TDI:

- *Nitration:* Reaction of toluene with nitric acid and a catalyst to form dinitrotoluene
- *Hydrogenation:* Reaction of the dinitrotoluene with hydrogen and a catalyst to form a mixture of isomers of toluene diamine (TDA)
- *Purification:* Distillation of the TDA mixture to produce meta-TDA
- *Phosgenation:* Reaction of the meta-TDA with phosgene to form a crude TDI mixture
- *Purification:* Distillation of the crude TDI mixture to produce an 80:20 mixture of 2,4-TDI and 2,6-TDI, known as TDI (80/20)
- *Differentiation:* Separation of the TDI (80/20) to produce pure 2,4-TDI and a 65:35 mixture of 2,4-TDI and 2,6-TDI, known as TDI (65/35).

### *Chemistry*

Each of the isocyanate functional groups in TDI can react with a hydroxyl group to form a urethane linkage. 2,4-TDI is an asymmetrical molecule and thus has two isocyanate groups of different reactivity. The 4-position is approximately four times more reactive than the 2-position. 2,6-TDI is a symmetrical molecule and thus has two isocyanate groups of similar reactivity, similar to the 2-position on 2,4-TDI. However, since both isocyanate groups are attached to the same aromatic ring, reaction of one isocyanate group will cause a change in the reactivity of the second isocyanate group.

### *Hazards*

Exposure to TDI and its vapours should be avoided. It is a well-documented causative agent of asthma.

Information on handling, personal protective equipment, exposure monitoring, transport, storage, sampling and analysis of TDI, dealing with accidents, and health and environmental information have been published

TDI is one of eleven statutorily listed "Extremely Hazardous Substances" under the New Jersey Toxic Catastrophe Prevention Act whose release "would produce a significant likelihood that persons exposed will suffer acute health effects resulting in death or permanent disability." N.J.S.A. 13:1K-21(e).

## Benzoic Acid

Benzoic acid, $C_7H_6O_2$ (or $C_6H_5COOH$), is a colorless crystalline solid and the simplest aromatic carboxylic acid. The name derived from gum benzoin, which was for a long time the only source for benzoic acid. Its salts are used as a food preservative and benzoic acid is an important precursor for the synthesis of many other organic substances. The salts and esters of benzoic acid are known as benzoates.

### *History*

Benzoic acid was discovered in the sixteenth century. The dry distillation of gum benzoin was first described by Nostradamus (1556), and subsequently by Alexius Pedemontanus (1560) and Blaise de Vigenère (1596).

Pioneer work in 1830 through a variety of experiences based on amygdalin, obtained from bitter almonds (the fruit of Prunus dulcis) oil by Pierre Robiquet and Antoine Boutron-Charlard, two French chemists, had produced benzaldehyde but they failed in working out a proper interpretation of the structure of amygdalin that would account for it, and thus missed the identification of the benzoyl radical $C_7H_5O$. This last step was achieved some few months later (1832) by Justus von Liebig and Friedrich Wöhler, who determined the composition of benzoic acid. These latter also investigated how hippuric acid is related to benzoic acid. In 1875 Salkowski discovered the antifungal abilities of benzoic acid, which was used for a long time in the preservation of benzoate-containing cloudberry fruits.

## Production

### Industrial Preparations

Benzoic acid is produced commercially by partial oxidation of toluene with oxygen. The process is catalysed by cobalt or manganese naphthenates. The process uses cheap raw materials, proceeds in high yield, and is considered environmentally green.

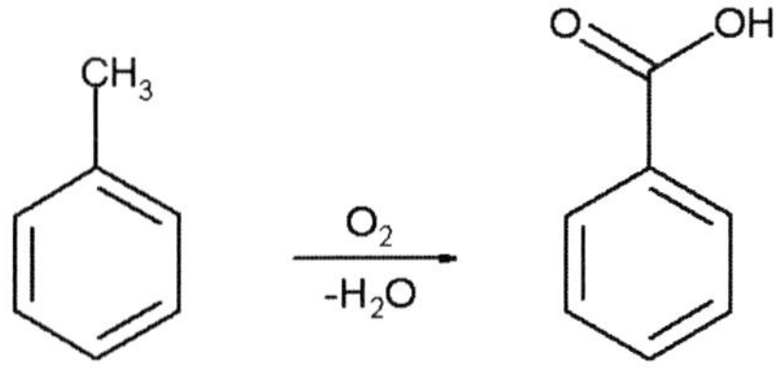


U.S. production capacity is estimated to be 126,000 tonnes per year (139,000 tons), much of which is consumed domestically to prepare other industrial chemicals.

### Laboratory Synthesis

Benzoic acid is cheap and readily available, so the laboratory synthesis of benzoic acid is mainly practiced for its pedagogical value. It is a common undergraduate preparation.

For all syntheses, benzoic acid can be purified by recrystallization from water because of its high solubility in hot water and poor solubility in cold water. The avoidance of organic solvents for the recrystallization makes this experiment particularly safe. Other possible recrystallization solvents include acetic acid (anhydrous or aqueous), benzene, acetone, petroleum ether, and a mixture of ethanol and water.

### By Hydrolysis

Like any other nitrile or amide, benzonitrile and benzamide can be hydrolyzed to benzoic acid or its conjugate base in acid or basic conditions.

### From Benzaldehyde

The base-induced disproportionation of benzaldehyde, the Cannizzaro reaction, affords equal amounts of benzoate and benzyl alcohol; the latter can be removed by distillation.

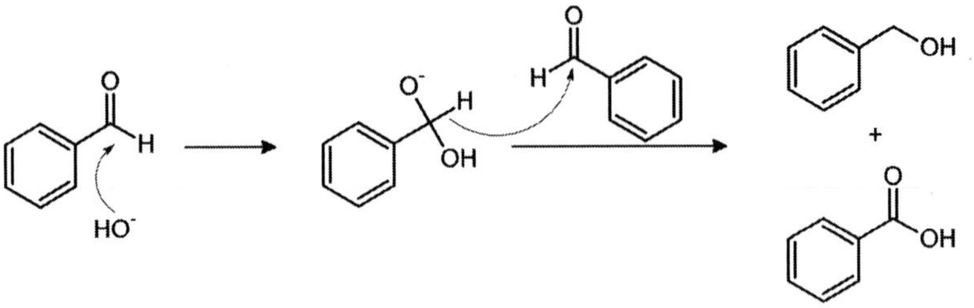


### From Bromobenzene

Bromobenzene can be converted to benzoic acid by "carbonation" of the intermediate phenylmagnesium bromide:

$$C_6H_5MgBr + CO_2 \rightarrow C_6H_5CO_2MgBr$$

$$C_6H_5CO_2MgBr + HCl \rightarrow C_6H_5CO_2H + MgBrCl$$

### From Benzyl Alcohol

Benzyl alcohol is refluxed with potassium permanganate or other oxidizing reagents in water.

The mixture is hot filtered to remove manganese dioxide and then allowed to cool to afford benzoic acid.

### From Benzyl Chloride

Benzoic acid can be prepared by oxidation of benzyl chloride in the presence of alkaline $KMnO_4$:

$$C_6H_5CH_2Cl + 2\ KOH + 2\ [O] \rightarrow C_6H_5COOK + KCl + H_2O$$

### Historical Preparation

The first industrial process involved the reaction of benzotrichloride (trichloromethyl benzene) with calcium hydroxide in water, using iron or iron salts as catalyst.

The resulting calcium benzoate is converted to benzoic acid with hydrochloric acid. The product contains significant amounts of chlorinated benzoic acid derivatives. For this reason, benzoic acid for human consumption was obtained by dry distillation of gum benzoin. Food-grade benzoic acid is now produced synthetically.

## Uses

### Calorimetry

Benzoic acid is the most commonly used chemical standard to determine the heat of capacity of a bomb calorimeter.

### Feedstock

Benzoic acid is used to make a large number of chemicals; important examples of which are:

- Benzoyl chloride, $C_6H_5C(O)Cl$, is obtained by treatment of benzoic with thionyl chloride, phosgene or one of the chlorides of phosphorus. $C_6H_5C(O)Cl$ is an important starting material for several benzoic acid derivates like benzyl benzoate, which is used in artificial flavours and insect repellents.
- Benzoate plasticizers, such as the glycol-, diethylengylcol-, and triethyleneglycol esters, are obtained by transesterification of methyl benzoate with the corresponding diol. Alternatively these species arise by treatment of benzoylchloride with the diol. These plasticizers are used similarly to those derived from terephthalic acid ester.
- Phenol, $C_6H_5OH$, is obtained by oxidative decarboxylation at 300-400°C. The temperature required can be lowered to 200°C by the addition of catalytic amounts of copper(II) salts. The phenol can be converted to cyclohexanol, which is a starting material for nylon synthesis.

### Food Preservative

Benzoic acid and its salts are used as a food preservatives, represented by the E-numbers E210, E211, E212, and E213. Benzoic acid inhibits the growth of mold, yeast and some bacteria. It is either added directly or created from reactions with its sodium, potassium, or calcium salt.

The mechanism starts with the absorption of benzoic acid in to the cell. If the intracellular pH changes to 5 or lower, the anaerobic fermentation of glucose through phosphofructokinase is decreased by 95%. The efficacy of benzoic acid and benzoate is thus dependent on the pH of the food. Acidic food and beverage like fruit juice (citric acid), sparkling drinks (carbon dioxide), soft drinks (phosphoric acid), pickles (vinegar) or other acidified food are preserved with benzoic acid and benzoates.

Typical levels of use for benzoic acid as a preservative in food are between 0.05–0.1%. Foods in which benzoic acid may be used and maximum levels for its application are laid down in international food law.

Concern has been expressed that benzoic acid and its salts may react with ascorbic acid (vitamin C) in some soft drinks, forming small quantities of benzene.

### *Medicinal*

Benzoic acid is a constituent of Whitfield's ointment which is used for the treatment of fungal skin diseases such as tinea, ringworm, and athlete's foot. As the principal component of benzoin resin, benzoic acid is also a major ingredient in both tincture of benzoin and Friar's balsam. Such products have a long history of use as topical antiseptics and inhalant decongestants.

### *Biology and Health Effects*

Benzoic acid occurs naturally free and bound as benzoic acid esters in many plant and animal species. Appreciable amounts have been found in most berries (around 0.05%). Ripe fruits of several *Vaccinium* species (e.g., cranberry, *V. vitis idaea*; bilberry, *V. macrocarpon*) contain as much as 0.03-0.13% free benzoic acid. Benzoic acid is also formed in apples after infection with the fungus *Nectria galligena*. Among animals, benzoic acid has been identified primarily in omnivorous or phytophageous species, e.g., in viscera and muscles of the Rock Ptarmigan (*Lagopus muta*) as well as in gland secretions of male muskoxen (*Ovibos moschatus*) or Asian bull elephants (*Elephas maximus*).

Gum benzoin contains up to 20% of benzoic acid and 40% benzoic acid esters.

Benzoic acid is present as part of hippuric acid (*N*-benzoylglycine) in urine of mammals, especially herbivores (Gr. *hippos* = horse; *ouron* = urine). Humans produce about 0.44 g/L hippuric acid per day in their urine, and if the person is exposed to toluene or benzoic acid it can rise above that level.

For humans, the World Health Organization's International Programme on Chemical Safety (IPCS) suggests a provisional tolerable intake would be 5 mg/kg body weight per day. Cats have a significantly lower tolerance against benzoic acid and its salts than rats and mice.

Lethal dose for cats can be as low as 300 mg/kg body weight. The oral $LD_{50}$ for rats is 3040 mg/kg, for mice it is 1940–2263 mg/kg.

In Taipei, Taiwan, a city health survey in 2010 found 30% of tested dried and pickled food products failed a test having too much benzoic acid, which is known to affect the liver and kidney, along with more serious issues like excessive cyclamate.

Benzoic acid activates bromelain.

## Chemistry

Reactions of benzoic acid can occur at either the aromatic ring or the carboxyl group:

### Aromatic Ring

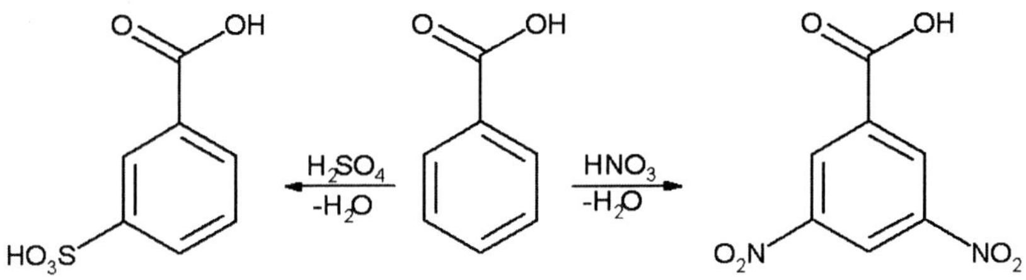


Electrophilic aromatic substitution reaction will take place mainly in 3-position due to the electron-withdrawing carboxylic group; i.e. benzoic acid is *meta* directing.

The second substitution reaction (on the right) is slower because the first nitro group is deactivating. Conversely, if an activating group (electron-donating) was introduced (e.g., alkyl), a second substitution reaction would occur more readily than the first and the disubstituted product might not accumulate to a significant extent.

### Carboxyl Group

All the reactions mentioned for carboxylic acids are also possible for benzoic acid.

- Benzoic acid esters are the product of the acid catalysed reaction with alcohols.
- Benzoic acid amides are more easily available by using activated acid derivatives (such as benzoyl chloride) or by coupling reagents used in peptide synthesis like DCC and DMAP.
- The more active benzoic anhydride is formed by dehydration using acetic anhydride or phosphorus pentoxide.

- Highly reactive acid derivatives such as acid halides are easily obtained by mixing with halogenation agents like phosphorus chlorides or thionyl chloride.
- Orthoesters can be obtained by the reaction of alcohols under acidic water free conditions with benzonitrile.
- Reduction to benzaldehyde and benzyl alcohol is possible using DIBAL-H, $LiAlH_4$ or sodium borohydride.
- The copper catalysed decarboxylation of benzoate to benzene may be effected by heating in quinoline. Also, Hunsdiecker decarboxylation can be achieved by forming the silver salt and heating. Benzoic acid can also be decarboxylated by heating with an alkali hydroxide or calcium hydroxide.

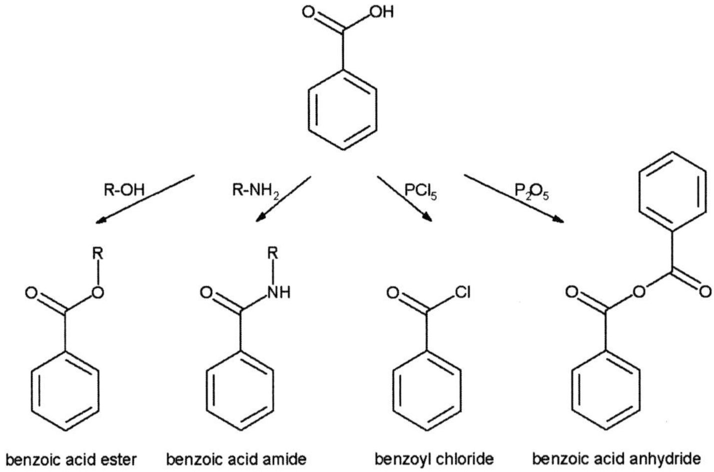


## Xylene

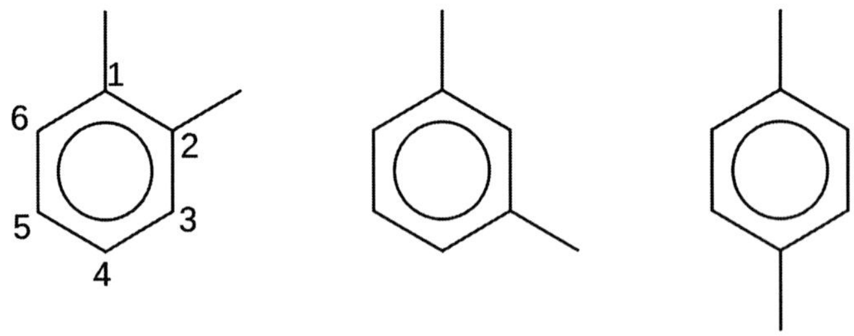


1,2-dimethylbenzene (xylene)*ortho*- 1,3-dimethylbenzene (xylene)*meta*- 1,4-dimethylbenzene (xylene)*para*-

*The Xylene Isomers*

Xylene encompasses three isomers of dimethylbenzene. The isomers are distinguished by the designations *ortho*- (*o*-), *meta*-

(*m*-), and *para*- (*p*-), which specify to which carbon atoms (of the benzene ring) the two methyl groups are attached. Counting the carbon atoms from one of the ring carbons bonded to a methyl group, and counting towards the second ring carbon bonded to a methyl group, the *o*- isomer has the IUPAC name of 1,2-dimethylbenzene, the *m*- isomer has the IUPAC name of 1,3-dimethylbenzene, and the *p*-isomer has the IUPAC name of 1,4-dimethylbenzene. The mixture is a slightly greasy, colourless liquid commonly encountered as a solvent. Several million tons are produced annually.

## Production

Xylenes represent about 0.5–1% of crude oil, depending on the source (hence xylenes are found in small amounts in gasoline and airplane fuels). It is mainly produced from reformate. It is also obtained from coal carbonisation derived from coke ovens. It is produced by dehydrocyclodimerization and by methylating of toluene and benzene.

Via the Isomar process, the ratio of isomers can be shifted to favour p-xylene, which is most valued. This conversion is catalysed by zeolites.

## Properties

Some chemical and physical properties differ from isomer to isomer. The melting point ranges from “47.87 °C (“54.17 °F) (*m*-xylene) to 13.26 °C (55.87 °F) (*p*-xylene). The boiling point for each isomer is around 140 °C (284.00 °F). The density of each is around 0.87 g/mL (7.26 lb/U.S. gallon or 8.72 lb/imp gallon) and thus is less dense than water. Xylene in air can be smelled at 0.08 to 3.7 parts of xylene per million parts of air (ppm) and can begin to be tasted in water at 0.53 to 1.8 ppm.

Xylenes form azeotropes with water and a variety of alcohols. With water the azeotrope consists of 60% xylenes and boils at 90 °C. As with many alkylbenzene compounds, xylenes form complexes with various halocarbons. The complexes of different isomers often have dramatically different properties from each other.

## Applications

### Terephthalic Acid and Related Derivatives

*p*-Xylene is the principal precursor to terephthalic acid and dimethyl terephthalate, both monomers used in the production of polyethylene terephthalate (PET) plastic bottles and polyester clothing.

98% of p-xylene production, and half of all xylene, is consumed in this way. *o*-Xylene is an important precursor to phthalic anhydride. The demand for isophthalic acid is relatively modest so *m*-xylene is rarely sought (and hence the utility of its conversion to the *o*- and *p*-isomers).

### *Solvent Applications*

Xylene is used as a solvent. In this application, the mixture of isomers is often referred to as xylenes or xylol. Solvent xylene often contains a small percentage of ethylbenzene. Like the individual isomers, the mixture is colorless, sweet-smelling, and highly flammable. Areas of application include printing, rubber, and leather industries. It is a common component of ink, rubber, adhesive, and leather industries. In thinning paints and varnishes, it can be substituted for toluene where slower drying is desired. Similarly it is a cleaning agent, e.g., for steel, silicon wafers, and chips.

### *Laboratory Uses*

It is used in the laboratory to make baths with dry ice to cool reaction vessels, and as a solvent to remove synthetic immersion oil from the microscope objective in light microscopy. In histology, xylene is also used for clearing the tissues following dehydration in preparation for paraffin wax infiltration. It is also used after sections have been stained to make them hydrophobic so that a coverslip may be applied with a resin in solvent.

### *Precursor to other Compounds*

Although conversion to terephthalic acid is the dominant chemical conversion, xylenes are precursors to other chemical compounds. For instance chlorination of both methyl groups gives the corresponding xylene dichlorides (bis(chloromethyl)benzenes).

### *Health Effects*

Xylenes are not highly toxic as indicated by the high values of the LD50, which range from 200 to 4000 mg/kg for animals. The principal mechanism of detoxification is oxidation to methylbenzoic acid and hydroxylation to hydroxylene.

## o-Xylene

*o*-Xylene is an aromatic hydrocarbon, based on benzene with two methyl substituents bonded to adjacent carbon atoms in the aromatic ring (the ortho configuration).

It is a constitutional isomer of *m*-xylene and *p*-xylene.

*o*-Xylene is largely used in the production of phthalic anhydride, and is generally extracted by distillation from a mixed xylene stream in a plant primarily designed for *p*-xylene production. Its main production is in Buenos Aires.

## Phthalic Anhydride

Phthalic anhydride is the organic compound with the formula $C_6H_4(CO)_2O$. It is the anhydride of phthalic acid. This colourless solid is an important industrial chemical, especially for the large-scale production of plasticizers for plastics. In 2002, approximately 4.6 billion kilograms were produced.

### *Synthesis and Production*

Phthalic anhydride was first reported in 1836 by Laurent. It is presently obtained by catalytic oxidation of *ortho*-xylene and naphthalene (Gibbs phthalic anhydride process):

$$C_6H_4(CH_3)_2 + 3\ O_2 \rightarrow C_6H_4(CO)_2O + 3\ H_2O$$

$$C_{10}H_8 + 4.5\ O_2 \rightarrow C_6H_4(CO)_2O + 2\ H_2O + 2\ CO_2$$

When separating the phthalic anhydride from byproducts such as *o*-xylene in water, or maleic anhydride, a series of "switch condensers" is required. Phthalic anhydride can also prepared from phthalic acid:

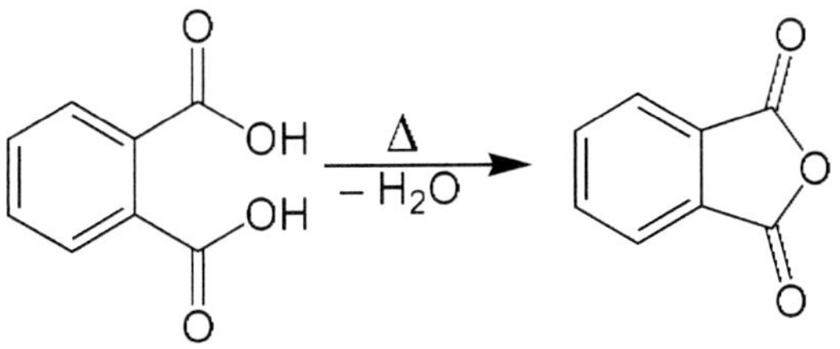


### *Applications in Industry and Organic Synthesis*

Phthalic anhydride is a versatile intermediate in organic chemistry, in part because it is bifunctional and cheaply available. It undergoes hydrolysis and alcoholysis. Hydrolysis by hot water forms *ortho*-phthalic acid. This process is reversible: Phthalic anhydride re-forms upon heating the acid above 180 °C. Hydrolysis of anhydrides is not typically a reversible process. However, phthalic acid is easily dehydrated to form phthalic anhydride due to the creation of a thermodynamically favourable 5-membered ring.

### Preparation of Phthalate Esters

As with other anhydrides, the alcoholysis reaction is the basis of the manufacture of phthalate esters, which are widely used (and controversial) plasticizers. In the 1980s, approximately $6.5\times10^{9}$ kg of these esters were produced annually, and the scale of production was increasing each year, all from phthalic anhydride. The process begins with the reaction of phthalic anhydride with alcohols, giving the monoesters:

$$C_6H_4(CO)_2O + ROH \rightarrow C_6H_4(CO_2H)CO_2R$$

The second esterification is more difficult and requires removal of water:

$$C_6H_4(CO_2H)CO_2R + ROH\ C_6H_4(CO_2R)_2 + H_2O$$

The most important diester is bis(2-ethylhexyl) phthalate ("DEHP"), used in the manufacture of polyvinyl chloride.

### Organic Synthesis

Phthalic anhydride is a precursor to a variety of reagents useful in organic synthesis. Important derivatives include phthalimide and its many derivatives. Chiral alcohols form half-esters, and these derivatives are often resolvable because they form diastereomeric salts with chiral amines such as brucine. A related ring-opening reaction involves peroxides to give the useful peroxy acid:

$$C_6H_4(CO)_2O + H_2O_2 \rightarrow C_6H_4(CO_3H)CO_2H$$

### Precursor to Dyestuffs

Phthalic anhydride is widely used in industry for the production of certain dyes. A well-known application of this reactivity is the preparation of the anthroquinone dye quinizarin by reaction with para-chlorophenol followed by hydrolysis of the chloride.

## p-Xylene

*p*-Xylene is an aromatic hydrocarbon, based on benzene with two methyl substituents. The "p" stands for *para*, identifying the location of the methyl groups as across from one another.

It is an isomer of xylene. Other isomers include *o*-xylene and *m*-xylene. *p*-Xylene is used on a large scale for the manufacture of terephthalic acid for polyester. Its polymer is known as Parylene. *p*-Xylene is produced by catalytic reforming of naphtha (a petroleum derivative) and separated in a series of distillation, adsorption or crystallization and reaction processes from *m*-xylene, *o*-xylene and ethylbenzene. Its melting point is the highest among this series of isomers, but simple crystallization does not allow easy purification due to the formation of eutectic mixtures.

## Dimethyl Terephthalate

Dimethyl terephthalate (DMT) is an organic compound with the formula $C_6H_4(CO_2CH_3)_2$. It is the diester formed from terephthalic acid and methanol. It is a white solid that melts to give a distillable colourless liquid.

### *Production*

DMT has been produced in a number of ways. Conventionally and still of commercial value is the direct esterification of terephthalic acid. Alternatively, it can be prepared by alternating oxidation and methyl-esterification steps from p-xylene via methylp-toluate.

### *Use*

DMT is used in the production of polyesters, including polyethylene terephthalate (PET) and polytrimethylene terephthalate. It consists of benzene substituted with carboxymethyl groups ($CO_2CH_3$) at the 1 and 4 positions. Because DMT is volatile, it is an intermediate is some schemes for the recyclic of PET, e.g. from plastic bottles.

Hydrogenation of DMT affords the diol cyclohexanedimethanol, which is a useful monomer.

# 6

# Polyethylene Terephthalate

Polyethylene terephthalate (sometimes written poly(ethylene terephthalate)), commonly abbreviated PET, PETE, or the obsolete PETP or PET-P, is a thermoplastic polymer resin of the polyester family and is used in synthetic fibres; beverage, food and other liquid containers; thermoforming applications; and engineering resins often in combination with glass fibre. The term 'polyethylene terephthalate' is a source of confusion because this substance, PET, does not contain polyethylene. Thus, the alternate form, 'poly(ethylene terephthalate)' is often used in scholarly journals for the sake of accuracy and clarity.

Depending on its processing and thermal history, polyethylene terephthalate may exist both as an amorphous (transparent) and as a semi-crystalline polymer. The semicrystalline material might appear transparent (particle size < 500 nm) or opaque and white (particle size up to a few microns) depending on its crystal structure and particle size. Its monomer (bis-β-hydroxyterephthalate) can be synthesized by the esterification reaction between terephthalic acid and ethylene glycol with water as a byproduct, or by transesterification reaction between ethylene glycol and dimethyl terephthalate with methanol as a byproduct. Polymerization is through a polycondensation reaction of the monomers (done immediately after esterification/ transesterification) with water as the byproduct.

The majority of the world's PET production is for synthetic fibres (in excess of 60%) with bottle production accounting for around 30% of global demand. In discussing textile applications, PET is generally referred to as simply "polyester" while "PET" is used most often to refer to packaging applications. The polyester industry makes up about 18% of world polymer production and is third after polyethylene (PE) and polypropylene (PP).

PET consists of polymerized units of the monomer ethylene terephthalate, with repeating $C_{10}H_8O_4$ units. PET is commonly recycled, and has the number "1" as its recycling symbol.

*Table 1 : PET*

| | |
|---|---|
| Young's modulus (E) | 2800–3100 MPa |
| Tensile strength($\sigma_t$) | 55–75 MPa |
| Elastic limit | 50–150% |
| notch test | 3.6 kJ/m$^2$ |
| Glass temperature | 75 °C |
| Vicat B | 170 °C |
| linear expansion coefficient ($\alpha$) | $7\times10^{-5}$/K |
| Water absorption (ASTM) | 0.16 |

### Uses

Because PET is an excellent barrier material, plastic bottles made from PET are widely used for soft drinks. For certain specialty bottles, PET sandwiches an additional polyvinyl alcohol to further reduce its oxygen permeability.

Biaxially oriented PET film (often known by one of its trade names, "Mylar") can be aluminized by evaporating a thin film of metal onto it to reduce its permeability, and to make it reflective and opaque (MPET). These properties are useful in many applications, including flexible food packaging and thermal insulation, such as "space blankets". Because of its high mechanical strength, PET film is often used in tape applications, such as the carrier for magnetic tape or backing for pressure sensitive adhesive tapes.

Non-oriented PET sheet can be thermoformed to make packaging trays and blisters. If crystallizable PET is used, the trays can be used for frozen dinners, since they withstand both freezing and oven baking temperatures. When filled with glass particles or fibres, it becomes significantly stiffer and more durable.

### History

PET was patented in 1941 by John Rex Whinfield, James Tennant Dickson and their employer the Calico Printers' Association of Manchester. The PET bottle was patented in 1973 by Nathaniel Wyeth. PET is used as substrate in thin film and solar cell

### Physical Properties

PET can be semi-rigid to rigid, depending on its thickness, and it is very lightweight. It makes a good gas and fair moisture barrier, as well as a good barrier to alcohol (requires additional "barrier" treatment) and solvents. It is strong and impact-resistant. It is naturally colorless with a high transparency.

### Intrinsic Viscosity

The intrinsic viscosity of the material, measured in deciliters per gram (d–l/g) is dependent upon the length of its polymer chains. The longer the polymer chains, the more entanglements between chains and therefore the higher the viscosity. The average chain length of a particular batch of resin can be controlled during polycondensation.

### The Intrinsic Viscosity Range of PET

Fibre grade

0.40 – 0.70 d–l/g Textile

0.72 – 0.98 d–l/g Technical, tire cord

Film grade

0.60 – 0.70 d–l/g BoPET (biaxially oriented PET film)

0.70 – 1.00 d–l/g Sheet grade for thermoforming

Bottle grade

0.70 – 0.78 d–l/g Water bottles (flat)

0.78 – 0.85 d–l/g Carbonated soft drink grade

Monofilament, engineering plastic

1.00 – 2.00 d–l/g

### Drying

PET is hygroscopic, meaning that it absorbs water from its surroundings. However, when this 'damp' PET is then heated, the water hydrolyzes the PET, decreasing its resilience. Thus, before the resin can be processed in a molding machine, it must be dried. Drying is achieved through the use of a desiccant or dryers before the PET is fed into the processing equipment.

Inside the dryer, hot dry air is pumped into the bottom of the hopper containing the resin so that it flows up through the pellets, removing moisture on its way. The hot wet air leaves the top of the hopper and is first run through an after-cooler, because it is easier to remove moisture from cold air than hot air. The resulting cool wet air is then passed through a desiccant bed. Finally the cool dry air leaving the desiccant bed is re-heated in a process heater and sent back through the same processes in a closed loop. Typically, residual moisture levels in the resin must be less than 5 parts per million (parts of water per million parts of resin, by weight) before processing. Dryer residence time should not be shorter than about four hours. This is because drying the material in less than 4 hours would require a temperature above 160 °C, at which level hydrolysis would begin inside the pellets before they could be dried out.

PET can also be dried in compressed air resin dryers. Compressed air dryers do not reuse drying air. Dry, heated compressed air is circulated through the PET pellets as in the desiccant dryer, then released to the atmosphere.

## *Copolymers*

In addition to pure (homopolymer) PET, PET modified by copolymerization is also available.

In some cases, the modified properties of copolymer are more desirable for a particular application. For example, cyclohexane dimethanol (CHDM) can be added to the polymer backbone in place of ethylene glycol. Since this building block is much larger (6 additional carbon atoms) than the ethylene glycol unit it replaces, it does not fit in with the neighboring chains the way an ethylene glycol unit would. This interferes with crystallization and lowers the polymer's melting temperature. Such PET is generally known as PETG (Eastman Chemical and SK Chemicals are the only two manufacturers). PETG is a clear amorphous thermoplastic that can be injection molded or sheet extruded. It can be coloured during processing.

Replacing terephthalic acid (right) with isophthalic acid (centre) creates a kink in the PET chain, interfering with crystallization and lowering the polymer's melting point.

Another common modifier is isophthalic acid, replacing some of the 1,4-(*para-*) linked terephthalate units. The 1,2-(*ortho-*) or 1,3-(*meta-*) linkage produces an angle in the chain, which also disturbs crystallinity.

Such copolymers are advantageous for certain molding applications, such as thermoforming, which is used for example to make tray or blister packaging from co-PET film, or amorphous PET sheet (A-PET) or PETG sheet. On the other hand, crystallization is important in other applications where mechanical and dimensional stability are important, such as seat belts. For PET bottles, the use of small amounts of isophthalic acid, CHDM, DEG or other comonomers can be useful: if only small amounts of comonomers are used, crystallization is slowed but not prevented entirely. As a result, bottles are obtainable via stretch blow molding ("SBM"), which are both clear and crystalline enough to be an adequate barrier to aromas and even gases, such as carbon dioxide in carbonated beverages.

### Crystals

Crystallization of polymers occurs when polymer chains fold up on themselves in a repeating, symmetrical pattern. Long polymer chains tend to become entangled on themselves, which prevents full crystallization in all but the most carefully controlled circumstances. PET is no exception to this rule; About 60% crystallization is the upper limit for commercial products, with the exception of polyester fibres. Besides, about 60% crystalline polymer about 40 % of the polymer chains remaining amorphous. Therefore PET is commonly called a semi-crystalline polymer.

PET in its natural state is a semi-crystalline resin. Clear products can be produced by rapidly cooling molten polymer below $T_g$ glass transition temperature to form an amorphous solid. Like glass, amorphous PET forms when its molecules are not given enough time to arrange themselves in an orderly, crystalline fashion as the melt is cooled. At room temperature the molecules are frozen in place, but if enough heat energy is put back into them by heating above $T_g$, they begin to move again, allowing crystals to nucleate and grow. This procedure is known as solid-state crystallization.

Like most materials, PET tends to produce spherulites containing many small crystallites when crystallized from an amorphous solid, rather than forming one large single crystal. Light tends to scatter as it crosses the boundaries between crystallites and the amorphous regions between them. This scattering means that crystalline PET is opaque and white in most cases. Fibre drawing is among the few industrial processes that produce a nearly single-crystal product.

### Degradation

PET is subject to various types of degradations during processing.

The main degradations that can occur are hydrolytic, thermal and, probably most important, thermal oxidation. When PET degrades, several things happen: discoloration, chain scissions resulting in reduced molecular weight, formation of acetaldehyde and cross-links ("gel" or "fish-eye" formation). Discoloration is due to the formation of various chromophoric systems following prolonged thermal treatment at elevated temperatures. This becomes a problem when the optical requirements of the polymer are very high, such as in packaging applications. The thermal and thermooxidative degradation results in poor processibility characteristics and performance of the material.

One way to alleviate this is to use a copolymer. Comonomers such as CHDM or isophthalic acid lower the melting temperature and reduce the degree of crystallinity of PET (especially important when the material is used for bottle manufacturing). Thus the resin can be plastically formed at lower temperatures and/or with lower force. This helps to prevent degradation, reducing the acetaldehyde content of the finished product to an acceptable (that is, unnoticeable) level. Other ways to improve the stability of the polymer is by using stabilizers, mainly antioxidants such as phosphites. Recently, molecular level stabilization of the material using nanostructured chemicals has also been considered.

### Acetaldehyde

Acetaldehyde is a colorless, volatile substance with a fruity smell. Although it forms naturally in some fruit, it can cause an off-taste in bottled water. Acetaldehyde forms by degradation of PET through the mishandling of the material. At high temperatures, (PET decomposes above 300 °C or 570 °F), high pressures, extruder speeds (excessive shear flow raises temperature) and long barrel residence times all contribute to the production of acetaldehyde. When acetaldehyde is produced, some of it remains dissolved in the walls of a container and then diffuses into the product stored inside, altering the taste and aroma. This is not such a problem for non-consumables (such as shampoo), for fruit juices (which already contain acetaldehyde), or for strong-tasting drinks like soft drinks. For bottled water, however, low acetaldehyde content is quite important, because if nothing masks the aroma, even extremely low concentrations (10–20 parts per billion in the water) of acetaldehyde can produce an off-taste.

### Antimony

Antimony (Sb) is a catalyst that is often used as antimony trioxide

($Sb_2O_3$) or antimony triacetate in the production of PET. After manufacturing a detectable amount of antimony can be found on the surface of the product. This residue can be removed with washing. Antimony also remains in the material itself and can thus migrate out into food and drinks. Exposing PET to boiling or microwaving can increase the levels of antimony significantly, possibly above USEPA maximum contamination levels. The drinking water limit in the USA for antimony is 6 parts per billion. Although antimony trioxide is of low toxicity when taken orally, its presence is still of concern. The Swiss Federal Office of Public Health investigated the amount of antimony migration, comparing waters bottled in PET and glass: the antimony concentrations of the water in PET bottles were higher, but still well below the allowed maximum concentration. The Swiss Federal Office of Public Health concluded that small amounts of antimony migrate from the PET into bottled water, but that the health risk of the resulting low concentrations is negligible (1% of the "tolerable daily intake" determined by the WHO). A later (2006) but more widely publicized study found similar amounts of antimony in water in PET bottles. The WHO has published a risk assessment for antimony in drinking water. Fruit juice concentrates (for which no guidelines are established), however, that were produced and bottled in PET in the UK were found to contain up to 44.7 µg/L of antimony, well above the EU limits for tap water of 5 µg/L.

### Toxicity of PET

Commentary published in Environmental Health Perspectives in April 2010 suggested that PET might yield endocrine disruptors under conditions of common use and recommended research on this topic. Proposed mechanisms include leaching of phthalates as well as leaching of antimony.

### Bottle Processing Equipment

There are two basic molding methods for PET bottles, one-step and two-step. In two-step molding, two separate machines are used. The first machine injection molds the preform, which resembles a test tube, with the bottle-cap threads already molded into place. The body of the tube is significantly thicker, as it will be inflated into its final shape in the second step using stretch blow molding.

In the second step, the preforms are heated rapidly and then inflated against a two-part mold to form them into the final shape of the bottle. Preforms (uninflated bottles) are now also used as containers for candy, and by some Red Cross chapters to distribute to homeowners to store medical history for emergency responders.

In one-step machines, the entire process from raw material to finished container is conducted within one machine, making it especially suitable for molding non-standard shapes (custom molding), including jars, flat oval, flask shapes etc. Its greatest merit is the reduction in space, product handling and energy, and far higher visual quality than can be achieved by the two-step system.

### *Polyester Recycling Industry*

While most thermoplastics can, in principle, be recycled, PET bottle recycling is more practical than many other plastic applications. The primary reason is that plastic carbonated soft drink bottles and water bottles are almost exclusively PET. PET has a resin identification code of 1. One of the uses for a recycled PET bottle is for the manufacture of polar fleece material. Among its many uses, companies such as English Retreads use the PET material to line their products. It can also make fibre for polyester products.

Because of the recyclability of PET and the relative abundance of post-consumer waste in the form of bottles, PET is rapidly gaining market share as a carpet fibre.

Mohawk Industries released everSTRAND in 1999, a 100% post-consumer recycled content PET fibre. Since that time, more than 17 billion bottles have been recycled into carpet fibre. Pharr Yarns, a supplier to numerous carpet manufacturers including Looptex, Dobbs Mills, and Berkshire Flooring, produces a BCF (bulk continuous filament) PET carpet fibre containing a minimum of 25% post-consumer recycled content.

PET, as with many plastics, is also an excellent candidate for thermal disposal (incineration), as it is composed of carbon, hydrogen, and oxygen, with only trace amounts of catalyst elements (but no sulfur). PET has the energy content of soft coal.

When recycling polyethylene terephthalate or PET or polyester, two ways generally have to be differentiated:

1. The chemical recycling back to the initial raw materials purified terephthalic acid (PTA) or dimethyl terephthalate (DMT) and ethylene glycol (EG) where the polymer structure is destroyed completely, or in process intermediates like bis-ß-hydroxyterephthalate

2. The mechanical recycling where the original polymer properties are being maintained or reconstituted.

Chemical recycling of PET will become cost-efficient only applying high capacity recycling lines of more than 50,000 tons/year. Such lines could only be seen, if at all, within the production sites of very large polyester producers. Several attempts of industrial magnitude to establish such chemical recycling plants have been made in the past but without resounding success. Even the promising chemical recycling in Japan has not become an industrial break through so far. The two reasons for this are: at first, the difficulty of consistent and continuous waste bottles sourcing in such a huge amount at one single site, and, at second, the steadily increased prices and price volatility of collected bottles. The prices of baled bottles increased for instance between the years 2000 and 2008 from about 50 Euro/ton to over 500 Euro/ton in 2008.

Mechanical recycling or direct circulation of PET in the polymeric state is operated in most diverse variants today. These kinds of processes are typical of small and medium-sized industry. Cost-efficiency can already be achieved with plant capacities within a range of 5 000 – 20 000 tons/year. In this case, nearly all kinds of recycled-material feedback into the material circulation are possible today. These diverse recycling processes are being discussed hereafter in detail.

Besides chemical contaminants and degradation products generated during first processing and usage, mechanical impurities are representing the main part of quality depreciating impurities in the recycling stream. Recycled materials are increasingly introduced into manufacturing processes, which were originally designed for new materials only. Therefore, efficient sorting, separation and cleaning processes become most important for high quality recycled polyester.

When talking about polyester recycling industry, we are concentrating mainly on recycling of PET bottles which are meanwhile used for all kinds of liquid packaging like water, carbonated soft drinks, juices, beer, sauces, detergents, household chemicals and so on. Bottles are easy to distinguish because of shape and consistency and separate from waste plastic streams either by automatic or hand sorting processes. The established polyester recycling industry consists of three major sections:

- PET bottle collection and waste separation—waste logistics

- Production of clean bottle flakes—flake production
- Conversion of PET flakes to final products—flake processing.

Intermediate product from the first section is baled bottle waste with a PET content greater than 90%. Most common trading form is the bale but also bricked or even loose, pre-cut bottles are common in the market. In the second section the collected bottles are converted to clean PET bottle flakes. This step can be more or less complex and complicated depending on required final flake quality. During third step PET bottle flakes are processed to any kind of products like film, bottles, fibre, filament, strapping or intermediates like pellets for further processing and engineering plastics.

Besides this external (post-consumer) polyester bottle recycling, numbers of internal (pre-consumer) recycling processes exist, where the wasted polymer material does not exit the production site to the free market, and instead is reused in the same production circuit. In this way, fibre waste is directly reused to produce fibre, preform waste is directly reused to produce preforms, and film waste is directly reused to produce film.

## PET Bottle Recycling

### *Purification and Decontamination*

The success of any recycling concept is hidden in the efficiency of purification and decontamination at the right place during processing and to the necessary or desired extent.

Generally, the following applies: the earlier in the process foreign substances are removed, and the more thoroughly this is done, the more efficient the process is.

The high plasticization temperature of PET in the range of 280°C is the reason why almost all common organic impurities such as PVC, PLA, polyolefin, chemical wood-pulp and paper fibres, polyvinyl acetate, melt adhesive, coloring agents, sugar and protein residues are transformed into coloured degradation products which, in their turn, might release reactive degradation products additionally. Then, the number of defects in the polymer chain increases considerably. Naturally, the particle size distribution of impurities is very wide, the big particles of 60–1000 μm—which are visible by naked eye and easy to filter—representing the lesser evil since their total surface is relatively small and the degradation speed is therefore lower. The influence of the microscopic particles, which—because they are many—increase the frequency of defects in the polymer, is comparable bigger.

The motto "What the eye does not see the heart cannot grieve over" is considered to be very important in many recycling processes. Therefore besides efficient sorting the removal of visible impurity particles by melt filtration processes is playing a particular part in this case.

In general, one can say that the processes to make PET bottle flakes from collected bottles are as versatile as the different waste streams are different in their composition and quality. In view of technology there isn't just one way to do it. There are meanwhile many engineering companies which are offering flake production plants and components, and it is difficult to decide for one or other plant design. Nevertheless there are processes which are sharing most of these principles. Depending on composition and impurity level of input material, the general following process steps are applied.

1. Bale opening, briquette opening
2. Sorting and selection for different colours, foreign polymers especially PVC, foreign matter, removal of film, paper, glass, sand, soil, stones and metals
3. Pre-washing without cutting
4. Coarse cutting dry or combined to pre-washing
5. Removal of stones, glass and metal
6. Air sifting to remove film, paper and labels
7. Grinding, dry and/or wet
8. Removal of low-density polymers (cups) by density differences
9. Hot wash
10. Caustic wash, and surface etching, maintaining intrinsic viscosity and decontamination
11. Rinsing
12. Clean water rinsing
13. Drying
14. Air sifting of flakes
15. Automatic flake sorting
16. Water circuit and water treatment technology
17. Flake quality control.

### *Impurities and material defects*

The number of possible impurities and material defects which accumulate in the polymeric material is increasing permanently—when processing as well as when using polymers—taking into account a growing service lifetime, growing final applications and repeated recycling. As far as recycled PET bottles are concerned, the defects mentioned can be sorted in the following groups:

1. Reactive polyester OH- or COOH- end groups are transformed into dead or non-reactive end groups, e.g. formation of vinyl ester end groups through dehydration or decarboxylation of terephthalate acid, reaction of the OH- or COOH- end groups with mono-functional degradation products like mono-carbonic acids or alcohols. Results are decreased reactivity during re-polycondensation or re-SSP and broadening the molecular weight distribution.
2. The end group proportion shifts toward the direction of the COOH end groups built up through a thermal and oxidative degradation. The results are decrease in reactivity, and increase in the acid autocatalytic decomposition during thermal treatment in presence of humidity.
3. Number of poly-functional macromolecules increases. Accumulation of gels and long-chain branching defects.
4. Number, concentration and variety of non polymer-identical organic and inorganic foreign substances are increasing. With every new thermal stress, the organic foreign substances will react by decomposition. This is causing the liberation of further degradation-supporting substances and coloring substances.
5. Hydroxide and peroxide groups build up at the surface of the products made of polyester in presence of air (oxygen) and humidity. This process is accelerated by ultraviolet light. During an ulterior treatment process, hydro peroxides are a source of oxygen radicals which are source of oxidative degradation. Destruction of hydro peroxides is to happen before the first thermal treatment or during plasticization and can be supported by suitable additives like antioxidants.

Taking into consideration the above-mentioned chemical defects and impurities, there is an ongoing modification of the following polymer characteristics during each recycling cycle, which are detectable by chemical and physical laboratory analysis.

In particular:

- Increase of COOH end groups
- Increase of colour number b
- Increase of haze (transparent products)
- Increase of oligomer content
- Reduction in filterability
- Increase of by-products content such as acetaldehyde, formaldehyde
- Increase of extractable foreign contaminants
- Decrease in colour L
- Decrease of intrinsic viscosity or dynamic viscosity
- Decrease of crystallization temperature and increase of crystallization speed
- Decrease of the mechanical properties like tensile strength, elongation at break or elastic modulus
- Broadening of molecular weight distribution.

The recycling of PET-bottles is meanwhile an industrial standard process which is offered by a wide variety of engineering companies.

### Processing Examples for Recycled Polyester

Recycling processes with polyester are almost as varied as the manufacturing processes based on primary pellets or melt. Depending on purity of the recycled materials polyester can be used today in most of the polyester manufacturing processes as blend with virgin polymer or increasingly as 100% recycled polymer. Some exceptions like BOPET-film of low thickness, special applications like optical film or yarns through FDY-spinning at > 6000 m/min or microfilaments and micro-fibres are produced from virgin polyester only.

### Simple Re-pelletizing of Bottle Flakes

This process consists in transforming bottle waste into flakes, by drying and crystallizing the flakes, by plasticizing and filtering, as well as by pelletizing. Product is an amorphous re-granulate of an intrinsic viscosity in the range of 0.55–0.7 d–!/g, depending on how complete pre-drying of PET flakes has been done.

Special feature are: acetaldehyde and oligomers are contained in the pellets at lower level; the viscosity is reduced somehow, the pellets are amorphous and have to be crystallized and dried before further processing.

Processing to:

- A-PET film for thermoforming,
- Addition to PET virgin production,
- BoPET packaging film,
- PET Bottle resin by SSP,
- Carpet yarn,
- Engineering plastic,
- Filaments,
- Non-woven,
- Packaging stripes,
- Staple fibre.

Choosing the re-pelletizing way means having an additional conversion process which is at the one side energy intensive, cost consuming and causes thermal destruction. At the other side the pelletizing step is providing the following advantages:

- Intensive melt filtration
- Intermediate quality control
- Modification by additives.
- Product selection and separation by quality
- Processing flexibility increased
- Quality uniformization.

### *Manufacture of PET-pellets or Flakes for Bottles (Bottle to Bottle) and A-PET*

This process is, in principle, similar to the one described above; however, the pellets produced are directly (continuously or discontinuously) crystallized and then subjected to a solid-state polycondensation (SSP) in a tumbling drier or a vertical tube reactor. During this processing step, the corresponding intrinsic viscosity of 0.80 – 0.085 d–!/g is rebuild again and, at the same time, the acetaldehyde content is reduced to < 1 ppm.

The fact that some machine manufacturers and line builders in Europe and USA make efforts to offer independent recycling processes, e.g. the so called bottle-to-bottle (B-2-B) process, such as BePET,

Starlinger, URRC or BÜHLER, aims at generally furnishing proof of the "existence" of the required extraction residues and of the removal of model contaminants according to FDA applying the so called challenge test, which is necessary for the application of the treated polyester in the food sector. Besides this process approval it is nevertheless necessary that any user of such processes has to constantly check the FDA-limits for the raw materials manufactured by himself for his process.

### Direct Conversion of Bottle Flakes

In order to save costs, an increasing number of polyester intermediate producers like spinning mills, strapping mills, or cast film mills are working on the direct use of the PET-flakes, from the treatment of used bottles, with a view to manufacturing an increasing number of polyester intermediates. For the adjustment of the necessary viscosity, besides an efficient drying of the flakes, it is possibly necessary to also reconstitute the viscosity through polycondensation in the melt phase or solid-state polycondensation of the flakes. The latest PET flake conversion processes are applying twin screw extruders, multi-screw extruders or multi-rotation systems and coincidental vacuum degassing to remove moisture and avoid flake pre-drying. These processes allow the conversion of undried PET flakes without substantial viscosity decrease caused by hydrolysis.

Looking at the consumption of PET bottle flakes, the main portion of about 70% is converted to fibres and filaments. When using directly secondary materials such as bottle flakes in spinning processes, there are a few processing principles to obtain.

High speed spinning processes for the manufacture of POY normally need a viscosity of 0.62–0.64 d–!/g. Starting from bottle flakes, the viscosity can be set via the degree of drying. The additional use of $TiO_2$ is necessary for full dull or semi dull yarn. In order to protect the spinnerets, an efficient filtration of the melt is, in any case is necessary. For the time being the amount of POY made of 100% recycling polyester is rather low because this process requires high purity of spinning melt. Most of the time a blend of virgin and recycled pellets is used.

Staple fibres are spun in an intrinsic viscosity range which rather lies somewhat lower and which should be between 0.58 and 0.62 d–!/g. In this case too, the required viscosity can be adjusted via drying or vacuum adjustment in case of vacuum extrusion. For adjusting the viscosity, however, an addition of chain length modifier like ethylene glycol or diethylene glycol can also be used.

Spinning non-woven — in the fine titer field for textile applications as well as heavy spinning non-woven as basic materials, e.g. for roof covers or in road building—can be manufactured by spinning bottle flakes. The spinning viscosity is again within a range of 0.58–0.65 d–!/g.

One field of increasing interest where recycled materials are used is the manufacture of high tenacity packaging stripes, and monofilaments. In both cases, the initial raw material is a mainly recycled material of higher intrinsic viscosity. High tenacity packaging stripes as well as monofilament are then manufactured in the melt spinning process.

## Recycling Back to the Initial Raw Materials

### *Glycolysis and Partial Glycolysis*

The polyester which has to be recycled is transformed into an oligomer by adding ethylene glycol or other glycols during thermal treatment. The aim and advantage of this way of processing is the possibility of separating the mechanical deposits directly and efficiently through a progressive and stepwise filtration. The filtration fineness of the last filtration step has a decisive effect on the quality of the end product. Taking partial recycling with partial glycolysis as an example, it is to be demonstrated how bottle waste can successfully be recycled in a continuously operating polyester line which is manufacturing pellets for bottle applications.

The task consists in feeding 10–25% bottle flakes while maintaining the quality of the bottle pellets which are manufactured on the line. This aim is solved by degrading the PET bottle flakes — already during their first plasticization which can be carried out in a single- or multi-screw extruder — to an intrinsic viscosity of about 0.30 d–!/g by adding small quantities of ethylene glycol and by subjecting the low viscosity melt stream to an efficient filtration directly after plasticization. Furthermore, temperature is brought to the lowest possible limit. In addition, with this way of processing, the possibility of a chemical decomposition of the hydro peroxides is possible by adding a corresponding P-stabilizer directly when plasticizing. The destruction of the hydro peroxide groups is, with other processes, already carried out during the last step of flake treatment for instance by adding $H_3PO_3$. The partially glycolyzed and finely filtered recycled material is continuously fed to the esterification or prepolycondensation reactor, the dosing quantities of the raw materials are being adjusted accordingly.

The treatment of polyester waste through total glycolysis to convert the polyester to bis-beta hydroxy-terephthalate, which is vacuum distilled and can be used, instead of DMT or PTA, as a raw material for polyester manufacture, has been executed on an industrial scale in Japan as experimental production.

### *Hydrolysis*

Recycling processes, through hydrolysis of the PET to PTA and MEG, are operating under high pressures under supercritical conditions. In this case, PET-waste will be directly hydrolyzed applying for instance supercritical water steam. Purification of crude terephthalic acid will be carried out by re-crystallization in acetic acid/water mixtures similar to PTA purification. Industrial-scale lines based on this chemistry have not been known to date.

### *Methanolysis*

Methanolysis is the recycling process which has been practiced and tested on a large scale for many years in the past. In this case, polyester waste is transformed with methanol into DMT, under pressure and in presence of catalysts. After this an efficient filtration of the methanolysis product is applied. Finally the crude DMT is purified by vacuum distillation. The methanolysis is only rarely carried out in industry today because polyester production based on DMT shrunk tremendously and with this DMT producers disappeared step by step during the last decade.

## Terephthalic Acid

Terephthalic acid is the organic compound with formula $C_6H_4(COOH)_2$. This colourless solid is a commodity chemical, used principally as a precursor to the polyester PET, used to make clothing and plastic bottles. Several billion kilograms are produced annually. It is one of three isomeric phthalic acids.

### *Properties*

Terephthalic acid is poorly soluble in water and alcohols, consequently up until around 1970 most crude terephthalic acid was converted to the dimethyl ester for purification. It sublimes when heated.

## Production

Terephthalic acid is produced by oxidation of *p*-xylene by oxygen in air:

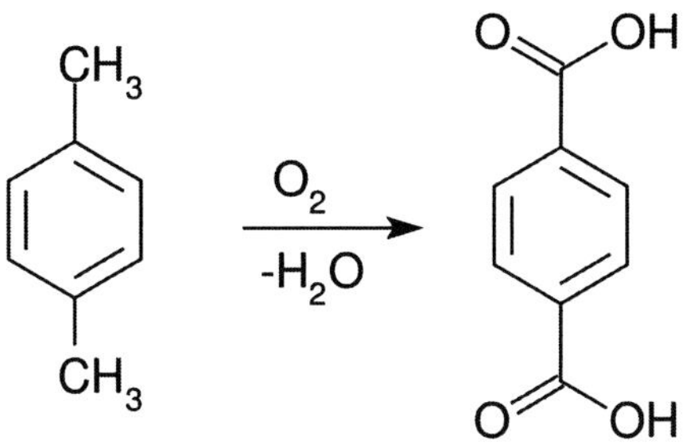


The oxidation is conducted using acetic acid as solvent and a catalyst composed of cobalt and manganese salts, using a bromide promoter. The yield is nearly quantitative. The most problematic impurity is 4-formylbenzoic acid, which is removed by hydrogenation of a hot aqueous solution. The solution is then cooled in a stepwise manner to crystallize highly pure terephthalic acid.

Alternatively, but not commercially significant, is the so-called "Henkel process" or "Raecke process," named after the company and patent holder, respectively. This process involves the rearrangement of phthalic acid to terephthalic acid via the corresponding potassium salts. Terephthalic acid can be prepared in the laboratory by oxidizing various para-disubstituted derivatives of benzene including Caraway Oil or a mixture of cymene and cuminol with chromic acid.

## Applications

Virtually the entire world's supply of terephthalic acid and dimethyl terephthalate are consumed as precursors to polyethylene terephthalate (PET). World production in 1970 was around 1.75 million tonnes. By 2006, global purified terephthalic acid (TPA) demand had exceeded 30 million tonnes.

There is a smaller, but nevertheless significant, demand for terephthalic acid in the production of polybutylene terephthalate and several other engineering polymers.

In the research laboratory, terephthalic acid has been popularized as a component for the synthesis of metal-organic frameworks.

The anxiolytic, analgesic, and antidepressant oxycodone occasionally comes as a terephthalate salt; however, the more usual salt of oxycodone is the hydrochloride. Pharmacologically, one milligramme of *terephthalas oxycodonae* is equivalent to 1.13 mgm of *hydrochloridum oxycodonae*.

## m-Xylene

*m*-Xylene is an aromatic hydrocarbon, based on benzene with two methyl substituents.

It is an isomer of *o*-xylene and *p*-xylene. The *m* stands for *meta*, meaning the two methyl substituents are at locants 1 and 3 on the aromatic ring.

The major chemical use of metaxylene is in the manufacture of isophthalic acid, which is used as a copolymer to alter the properties of polyethylene terephthalate (PET) making PET more suitable for the manufacture of soft drinks bottles. It is also used as a raw material in the manufacture of 2,4- and 2,6-xylidine as well as a range of smaller-volume chemicals.

## Isophthalic Acid

Isophthalic acid is an organic compound with the formula $C_6H_4(CO_2H)_2$. This colourless solid is an isomer of phthalic acid and terephthalic acid.

These aromatic dicarboxylic acids are used as precursors (in form of acyl chlorides) to commercially important polymers, e.g. the fire-resistant material Nomex. Mixed with terephthalic acid, isophthalic acid is used in the production of resins for drink bottles. The high-performance polymer polybenzimidazole is produced from isophthalic acid.

### *Preparation*

Isophthalic acid is produced on the billion kilogram/y scale by oxidizing meta-xylene using oxygen. The process employs a cobalt-manganese catalyst. In the laboratory, chromic acid can be used as the oxidant. It also arises by fusing potassium meta-sulphobenzoate, or meta-brombenzoate with potassium formate (terephthalic acid is also formed in the last case).

The barium salt (as its hexahydrate) is very soluble (a distinction between phthalic and terephthalic acids). Uvitic acid, 5-methylisophthalic acid, is obtained by oxidizing mesitylene or by condensing pyroracemic acid with baryta water.

## Alkyd

An alkyd is a polyester modified by the addition of fatty acids and other components. They are derived from polyols and a dicarboxylic acid or carboxylic acid anhydride. The term alkyd is a modification of the original name "alcid", reflecting the fact that they are derived from *alc*ohol and organic ac*ids*. The inclusion of the fatty acid confers a tendency to form flexible coating. Alkyds are used in paints and in moulds for casting. They are the dominant resin or "binder" in most commercial "oil-based" coatings. Approximately 200,000 tons of alkyd resins are produced each year.. The original alkyds were compounds of glycerol and phthalic acid sold under the name Glyptal. These were sold as substitutes for the Copal resins and made varnishes which were much paler in colour. From these, the alkyds we know today were developed.

### *Manufacture*

There are two types of alkyd resins, drying (including semi drying) and nondrying. Both types are typically produced from dicarboxylic acids or anhydrides, such as phthalic anhydride or maleic anhydride, and polyols, such as trimethylolpropane, glycerine, or pentaerythritol. Alkyds are also used in items such as resins and paints.

For the "drying" resins, triglycerides derived from polyunsaturated fatty acids (often derived from plant and vegetable oils, e.g. linseed oil). These drying alkyds are cured in air. The drying speed and the nature of the coatings depends on the amount and type of drying oil employed (more polyunsaturated oil means faster reaction in air) and use of metal salts, the so-called oil drying agents. These metal complex that catalyze crosslinking of the unsaturated sites.

Alkyd coatings are produced in two processes; fatty acid process and the alcoholysis or mono-glyceride process. Higher quality higher performance alkyds are produced in the fatty acid process where the composition of the resulting resin can be more precisely controlled. In this process an acid anhydride, a polyol and an unsaturated fatty acid are combined and cooked together until the product has achieved a predetermined level of viscosity, penta alkyds are made this way. More economical alkyd resins are produced from the alcoholysis or glyceride process where end product quality control is not as paramount. In this process raw vegetable oil, high in unsaturated component, is combined with additional polyol and heated to cause transesterification of the triglycerides into a mixture of mono- and diglyceride oils. To this resulting mixture acid anhydride is added to build molecular weight of the resin into roughly the same product as in the fatty acid process. However, the alcoholysis process, also known as the glyceride process, produces a more randomly oriented structure. To remove the water produced as a by-product and to increase the reaction rate, surplus phthalic anhydride was added. Water was thus removed with the unreacted acid by heating the bulk to a temperature needed to do this. It meant that the reaction was not as controllable as would be liked. A new process was introduced, in which Xylene was added to produce an azeotrope with the water. This gave greater control at a lower temperature, this also produced resins at a lower viscosity, useful in making high solids paints,this process was known as the AZO process. In both cases the resulting product is a polyester resin to which pendant drying oil groups are attached. At the conclusion of both processes the resin is purified, diluted in solvent and sold to paint and varnish makers.

### *Metalcasting*

Alkyd or oil-urethane binders are used in metalcasting for the creation of sand-based moulds. The alkyd resin is mixed with a polymeric isocyanate and a metallic drier, which speeds the reaction. Unlike other no-bake mould technologies, the process yields no toxic fumes but the moulds need a longer air curing.

### *Coating Types*

Alkyd coatings are typically sold in three classes: long, medium, and short. These terms represent the relative fraction of drying oil component in the resin. Long oil alkyds have a high percentage of drying oil content and are generally sold as medium duty coatings for the consumer market. Medium oil alkyds have less drying oil and have a higher percentage of large molecular weight polyester backbone. They dry slower and are employed as high gloss coatings and wood finishes. At the bottom end are short oil alkyds where the percentage of drying oil is very low in relation to the base polyester polymer or backbone chain. These coatings will not air dry or harden unless heated. Short oil alkyds are employed as baking enamels for finished metal products, in conjunction with amino/formaldehyde resins. Alkyds are also modified with phenolic resin, styrene, vinyl toluene, acrylic monomers (to make them dry more quickly), and polyurethanes. By adding certain modifiying resins, it is possible to produce thixotropic alkyds for decorative use. The latest alkyds are short oil A/D resins where the oil length is shortened by chain stopping with benzoic acid and now Para Tertiary Butyl Benzoic acid (Alkydal M 48), these have a better controlled molecular weight distribution and better durability. Alkyds for decorative use have extra oil cooked in to lengthen them and to make them more durable. Short oil resins used in stoving enamels are made from non-drying saturated oils or fatty acids. These usually have much higher hydroxyl and acid values, to be able to react with the hydroxyl groups of the amino resins. These mixtures are usually stabilized with amines to prevent gelling on storage.

Because the major components of an alkyd coating, i.e. fatty acids and triglyceride oils, are derived from low cost renewable resources, this has kept the cost of alkyd coatings very low despite ever increasing cost of petroleum, which is the predominant raw material source of most other coatings such as vinyls, acrylics, epoxies, and polyurethanes. Typical sources of drying oils for alkyd coatings are: linseed, tung, sunflower oil, safflower oil, walnut oil, soybean oil, fish oil, corn oil, DCO. (made by dehydrating castor oil, which creates a semi drying, conjugated, oil/fatty acid), and tall oil (resinous oil by-product from pulp and paper manufacturing). Non drying/plasticizer resins are made from castor, palm, coconut oils and cardura (a synthetic fatty, versatic acid). Dehydrated castor oil was at one time the only oil allowed to be used in resin manufacture in India, no edible oils were allowed to be used.

## Polyamide-imide

Polyamide-imides are thermoplastic amorphous polymers that have exceptional mechanical, thermal and chemical resistant properties. These properties put polyamide-imides at the top of the price and performance pyramid. Polyamide-imides are produced by Solvay Advanced Polymers under the trademark Torlon. Other high-performance polymers in this same realm are polyetheretherketones and polyimides.

Polyamide-imides hold, as the name suggests, a positive synergy of properties from both polyamides and polyimides, such as high strength, melt processibility, exceptional high heat capability, and broad chemical resistance. Polyamide-imide polymers can be processed into a wide variety of forms – from injection or compression molded parts and ingots – to coatings, films, fibres and adhesives. Generally these articles reach their maximum properties with a subsequent thermal cure process.

### Chemistry

The currently popular commercial methods to synthesize polyamide-imides are the acid chloride route and the isocyanate route.

### Acid Chloride Route

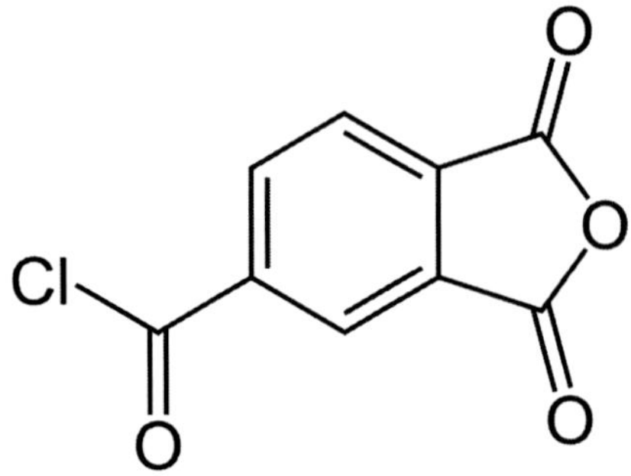

*Trimelletic acid chloride*

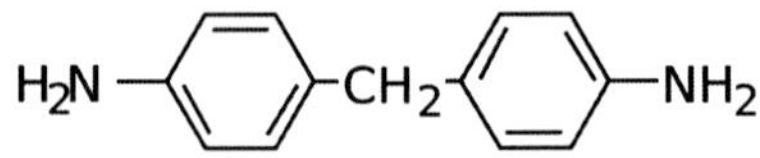

*Methylene dianiline*

The earliest route to polyamide-imides is the condensation of an aromatic diamine, such as methylene dianiline (MDA) and trimellitic acid chloride (TMAC). Reaction of the anhydride with the diamine produces an intermediate amic acid.

The acid chloride functionality reacts with the aromatic amine to give the amide bond and hydrochloric acid (HCl) as a by-product. In the commercial preparation of polyamideimides, the polymerization is carried out in a dipolar, aprotic solvent such as N-methylpyrrolidone (NMP), dimethylacetamide (DMAC), dimethylformamide (DMF), or dimethylsulfoxide (DMSO) at temperatures between 20-60°C.

The byproduct HCl must be neutralized in situ or removed by washing it from the precipitated polymer. Further thermal treatment of the polyamideimide polymer increases molecular weight and causes the amic acid groups to form imides with the evolution of water.

### Diisocyanate Route

This is the primary route to polyamide-imides which are used as wire enamels. A diisocyanate, often 4,4'- methylenediphenyl-diisocyanate (MDI), is reacted with trimellitic anhydride (TMA). The product achieved at the end of this process is a high molecular weight, fully imidized polymer solution with no condensation byproducts, since the carbon dioxide gas byproduct is easily removed.

This form is convenient for the manufacture of wire enamel or coatings. The solution viscosity is controlled by stoichiometry, monofunctional reagents, and polymer solids. The typical polymer solids level is 35-45% and it may be diluted further by the supplier or user with diluents.

### Fabrication

Polyamide-imides are commercially used for coatings and molded articles.

### Coatings

The product used mainly for coatings is sold in a powdered form and is roughly 50% imidized. One of the major uses is as a magnet wire enamel.

The magnet wire enamel is made by dissolving the PAI powder in a strong, aprotic solvent such as N-methyl pyrrolidone. Diluents and other additives can be added to provide the correct viscosity for application to the copper or aluminum conductor. Application is typically done by drawing the conductor through a bath of enamel and then through a die to control coating thickness. The wire is then passed through an oven to drive off the solvent and cure the coating. The wire usually is passed through the process several times to achieve the desired coating thickness.

The PAI enamel is very thermally stable as well as abrasion and chemical resistant. PAI is often used over polyester wire enamels to achieve higher thermal ratings.

PAI is also used in decorative, corrosion resistant coatings for industrial uses, often in conjunction with fluoropolymers. The PAI aids in adhering the fluoropolymer to the metal substrate. They also find usage in non-stick cookware coatings. While solvents can be used, some water-borne systems are used. These are possible because the amide-imide contains acid functionality.

### Molded or Machined Articles

The polyamide-imides used for molded articles are also based on aromatic diamines and trimellitic acid chloride, but the diamines are different from those used in the products used for coatings and the polymer is more fully imidized prior to compounding and pellitizing.

Resins for injection molding include unreinforced, glass-fibre reinforced, carbon fibre reinforced, and wear resistant grades.

These resins are sold at a relatively low molecular weight so they can be melt processed by extrusion or injection-molding. The molded articles are then thermally treated for several days at temperatures up to 260°C (500°F).

During this treatment, commonly referred to a postcure, the molecular weight increases through change extension and the polymer get much stronger and more chemically resistant. Prior to postcure, parts can be reground and reprocessed. After postcure, reprocessing is not practical.

## Properties of Molded PAI

### Wear-Resistant PAI Grades

| Property | Test Method | units | 4275 | 4301 | 4435 | 4630 | 4645 |
|---|---|---|---|---|---|---|---|
| Tensile Strength | ASTM D 638 | MPa (kpsi) | 117 (16.9) | 113 (16.4) | 94 (13.6) | 81 (11.8) | 114 (16.6) |
| Tensile Modulus | ASTM D 638 | GPa (kpsi) | 8.8 (1,280) | 6.8 (990) | 14.5 (2,100) | 7.4 (1,080) | 18.6 (2,700) |
| Tensile Elongation | ASTM D 638 | % | 2.6 | 3.3 | 1.0 | 1.9 | 0.8 |
| Flexural Strength | ASTM D 790 | MPa (kpsi) | 208 (30.2) | 215 (31.2) | 152 (22.0) | 131 (19.0) | 154 (22.4) |
| Flexural Modulus | ASTM D 790 | GPa (kpsi) | 7.3 (1.060) | 6.9 (1,000) | 14.8 (2,150) | 6.8 (990) | 12.4 (1,800) |
| Compressive Strength | ASTM D 695 | MPa (kpsi) | 123 (17.8) | 166 (24.1) | 138 (20.0) | 99 (14.4) | 157 (22.8) |
| Izod Impact Strength, Notched | ASTM D 256 | J/m (ft-lb/in) | 85 (1.6) | 64 (1.2) | 43 (0.8) | 48 (0.9) | 37 (0.7) |
| Izod Impact Strength, Unnotched | ASTM D 4812 | J/m (ft-lb/in) | 270 (5) | 430 (8) | 210 (4) | 160 (3) | 110 (2) |
| Heat Deflection Temperature at 264 psi | ASTM D 648 | °C (°F) | 280 (536) | 279 (534) | 278 (532) | 280 (536) | 281 (538) |
| Coefficient Linear Thermal Expansion | ASTM D 696 | ppm/°C (ppm/°F) | 25 (14) | 25 (14) | 14 (8) | 16 (9) | 9 (3) |

## High-strength Grades

| *Property* | *Test Method* | *units PAI* | *neat PAI* | *30% GF PAI* | *30% CF* |
|---|---|---|---|---|---|
| Tensile Strength | ASTM D 638 | MPa (kpsi) | 152 (22.0) | 221 (32.1) | 221 (32.0) |
| Tensile Modulus | ASTM D 638 | GPa (kpsi) | 4.5 (650) | 14.5 (2,110) | 16.5 (2,400) |
| Tensile Elongation | ASTM D 638 | % | 7.6 | 2.3 | 1.5 |
| Flexural Strength | ASTM D 790 | MPa (kpsi) | 241 (34.9) | 333 (48.3) | 350 (50.7) |
| Flexural Modulus | ASTM D 638 | GPa (kpsi) | 5.0 (730) | 11.7 (1,700) | 16.5 (2,400) |
| Compressive Strength | ASTM D 695 | MPa (kpsi) | 221 (32.1) | 264 (38.3) | 254 (36.9) |
| Shear Strength | ASTM D 732 | MPa (kpsi) | 128 (18.5) | 139 (20.1) | 119 (17.3) |
| Izod Impact Strength | ASTM D 256 | J/m (ftlb/in) | 144 (2.7) | 80 (1.5) | 48 (0.9) |
| Izod Impact Strength-Unnotched | ASTM D 4812 | J/m (ftlb/in) | 1070 (20) | 530 (10) | 320 (6) |
| Heat Deflection Temperature @ 264 psi | ASTM D 648 | °C (°F) | 278 (532) | 282 (540) | 282 (540) |
| Coefficient Linear Thermal Expansion | ASTM D 696 | ppm/°C (ppm/°F) | 31 (17) | 16 (9) | 9 (5) |
| Volume Resistivity | ASTM D 257 | ohm-cm | 2e17 | 2e17 | |
| Specific Gravity | ASTM D 792 | | 1.42 | 1.61 | 1.48 |
| Water Absorption, 24 hr | ASTM D 570 | % | 0.33 | 0.24 | 0.26 |

## Injection Molding Polyamide-imide

### Pre-drying the Resin

PAI resin is hygroscopic, and will pick up ambient moisture. Before processing PAI resin, drying is required to avoid brittle parts, foaming, and other molding problems. The resin must be dried to a moisture content of 500 ppm or less. A desiccant dryer capable of maintaining a dew point of -40°F (-40°C) is recommended. If drying is done in pans or trays, put the resin in layers no more than 2 to 3 inches (5 to 8 cm) deep in drying trays. Dry for 24 hours at 250°F, or 16 hours at 300°F, or 8 hours at 350°F. If drying at 350°F (177°C), limit drying time to 16 hours. For the injection molding press,a desiccant hopper dryer is recommended. The circulating air suction pipe should be at the base of the hopper, as near the feed throat as possible. During extended runs, keep the resin covered and re-dry if necessary.

### Injection Molding Equipment

In general, modern reciprocating-screw injection molding presses with microprocessor controls capable of closed-loop control are recommended for molding PAI. The press should be fitted with a low compression ratio, constant taper screw. The compression ratio should be between 1.1 and 1.5 to 1, and no check device should be used.

***Starting Point Temperatures***

| *Zone* | *Temp, °F* | *Temp, °C* |
|---|---|---|
| Feed Zone | 580 | 304 |
| Middle Zone | 620 | 327 |
| Front Zone | 650 | 343 |
| Nozzle | 700 | 371 |

The mold temperature should be in the range of 325°F to 425°F (163°C to 218°C).

## Polyester Resin

Polyester resins are unsaturated resins formed by the reaction of dibasic organic acids and polyhydric alcohols. Polyester resins are used in sheet moulding compound, bulk moulding compound and the toner of laser printers. Wall panels fabricated from polyester resins reinforced with fiberglass — so-called fiberglass reinforced plastic (FRP) — are typically used in restaurants, kitchens, restrooms and other areas that require washable low-maintenance walls.

Unsaturated polyesters are condensation polymers formed by the reaction of polyols (also known as polyhydric alcohols), organic compounds with multiple alcohol or hydroxy functional groups, with saturated or unsaturated dibasic acids. Typical polyols used are glycols such as ethylene glycol; acids used are phthalic acid and maleic acid. Water, a by-product of esterification reactions, is continuously removed, driving the reaction to completion. The use of unsaturated polyesters and additives such as styrene lowers the viscosity of the resin. The initially liquid resin is converted to a solid by cross-linking chains. This is done by creating free radicals at unsaturated bonds, which propagate in a chain reaction to other unsaturated bonds in adjacent molecules, linking them in the process. The initial free radicals are induced by adding a compound that easily decomposes into free radicals. This compound is usually and incorrectly known as the catalyst. Substances used are generally organic peroxides such as benzoyl peroxide or methyl ethyl ketone peroxide. Polyester resins are thermosetting and, as with other resins, cure exothermically. The use of excessive catalyst can, therefore, cause charring or even ignition during the curing process. Excessive catalyst may also cause the product to fracture or form a rubbery material.

# 7

# Petroleum Technology

## Hydraulic Fracturing

Hydraulic fracturing, often called fracking, fracing or hydrofracking, is the process of initiating and subsequently propagating a fracture in a rock layer, by means of a pressurized fluid, in order to release petroleum, natural gas, coal seam gas, or other substances for extraction. The fracturing, known colloquially as a frack job (or frac job), is done from a wellbore drilled into reservoir rock formations. The energy from the injection of a highly pressurized fluid, such as water, creates new channels in the rock which can increase the extraction rates and ultimate recovery of fossil fuels.

Hydraulic fractures may be natural or created by human activity, and are extended by internal fluid pressure which opens the fracture and causes it to extend through the rock. Natural hydraulic fractures include igneous dikes, sills and fracturing by ice as in frost weathering. Man-made fluid-driven fractures are formed at depth in a borehole and extend into targeted formations. The fracture width is typically maintained after the injection by introducing a *proppant* into the injected fluid. Proppant is a material, such as grains of sand, ceramic, or other particulates, that prevent the fractures from closing when the injection is stopped.

The practice of hydraulic fracturing has come under scrutiny internationally due to concerns about environmental and health safety, and has been suspended or banned in some countries.

### *Purpose*

The technique of hydraulic fracturing is used to increase or restore the rate at which fluids, such as oil or water, or natural gas can be produced from subterranean natural reservoirs, including unconventional reservoirs such as shale rock or coal beds. Hydraulic fracturing enables the production of natural gas and oil from rock formations deep below the earth's surface (generally 5,000-20,000 feet or 1,500-6,100 m). At such depth, there may not be sufficient porosity and permeability to allow natural gas and oil to flow from the rock into the wellbore at economic rates. Thus, creating conductive fractures in the rock is essential to extract gas from shale reservoirs because of the extremely low natural permeability of shale, which is measured in the microdarcy to nanodarcy range. Fractures provide a conductive path connecting a larger area of the reservoir to the well, thereby increasing the area from which natural gas and liquids can be recovered from the targeted formation.

While the main industrial use of hydraulic fracturing is in stimulating production from oil and gas wells, hydraulic fracturing is also applied to:

- Stimulating groundwater wells
- Preconditioning rock for caving or inducing rock to cave in mining
- As a means of enhancing waste remediation processes, usually hydrocarbon waste or spills
- Dispose of waste by injection into deep rock formations
- As a method to measure the stress in the earth
- For heat extraction to produce electricity in an Enhanced Geothermal System.

### Method

A hydraulic fracture is formed by pumping the fracturing fluid into the wellbore at a rate sufficient to increase pressure downhole to exceed that of the fracture gradient of the rock. The rock cracks and the fracture fluid continues farther into the rock, extending the crack still farther, and so on. To keep this fracture open after the injection stops, a solid proppant, commonly a sieved round sand, is added to the fluid. The propped fracture is permeable enough to allow the flow of formation fluids to the well. Formation fluids include gas, oil, salt water, fresh water and fluids introduced to the formation during completion of the well during fracturing.

The location of fracturing along the length of the borehole can be controlled by inserting composite plugs, also known as bridge plugs, above and below the region to be fractured. This allows a borehole to be progressively fractured along the length of the bore without fluid leaking through previously fractured regions. Fluid and proppant are introduced to the working region through piping in the upper plug. This method is commonly referred to as "plug and perf."

Typically, hydraulic fracturing is performed in cased wellbores and the zones to be fractured are accessed by perforating the casing at those locations.

### *Well Types*

While hydraulic fracturing is many times performed in vertical wells, today it is more often performed in horizontal wells. Horizontal drilling involves wellbores where the terminal drillhole is completed as a 'lateral' that extends parallel with the rock layer containing the substance to be extracted. For example, laterals extend 1,500 to 5,000 feet in the Barnett Shale basin. In contrast, a vertical well only accesses the thickness of the rock layer, typically 50–300 feet. Horizontal drilling also reduces surface disruptions as fewer wells are required. Drilling a wellbore produces rock chips and fine rock particles that may enter cracks and pore space at the wellbore wall, reducing the permeability at and near the wellbore. This reduces flow into the borehole from the surrounding rock formation, and partially seals off the borehole from the surrounding rock. Hydraulic fracturing can be used to restore permeability.

Hydraulic fracturing is commonly applied to wells drilled in low permeability reservoir rock. An estimated 90 percent of the natural gas wells in the United States use hydraulic fracturing to produce gas at economic rates.

### *Fracturing*

The fluid injected into the rock is typically a slurry of water, proppants, and chemical additives. Additionally, gels, foams, and compressed gases, including nitrogen, carbon dioxide and air can be injected. Various types of proppant include silica sand, resin-coated sand, and man-made ceramics. These vary depending on the type of permeability or grain strength needed. Sand containing naturally radioactive minerals is sometimes used so that the fracture trace along the wellbore can be measured. Chemical additives are applied to tailor the injected material to the specific geological situation, protect the well, and improve its operation, though the injected fluid is approximately more than 99 percent water and proppant, varying slightly based on the type of well. The composition of injected fluid is changed during the operation of a well over time, that is initially acid is used to increase permeability, then proppants are used with a gradual increase in their size and/or density, and at the end the well is flushed with water under pressure. Injected fluid is to some degree recovered and stored in pits or containers. Although the concentrations of the chemical additives are very low, the recovered fluid may be harmful due in part to hydrocarbons picked up from the formation. The recovered fluid is often processed so that part of it can be reused in another fracking operation, part released into the environment after treatment, and some left in the geologic formation.

Hydraulic fracturing equipment used in oil and natural gas fields usually consists of a slurry blender, one or more high pressure, high volume fracturing pumps (typically powerful triplex, or quintiplex pumps) and a monitoring unit. Associated equipment includes fracturing tanks, high pressure treating iron, a chemical additive unit (used to accurately monitor chemical addition) low pressure pipes and gauges for flow rate, fluid density, and treating pressure. Fracturing equipment operates over a range of pressures and injection rates, and can reach up to 100 MPa (15,000 psi) and 265 L/s (100 barrels per minute).

### Fracture Monitoring

Microseismic monitoring is commonly used to estimate the size and orientation of hydraulically induced fractures. Microseismic activity is measured by placing an array of geophones in a nearby wellbore. By mapping the location of any small seismic events associated with the growing hydraulic fracture, the approximate geometry of the fracture is inferred. Tiltmeter arrays, deployed on the surface or down a well, provide another technology for monitoring the strains produced by hydraulic fracturing.

Emission of gases displaced by hydraulic fracturing into the atmosphere may be detected via atmospheric gas monitoring, and can be quantified directly via the eddy covariance flux measurements.

### Completion

Advances in completion technology have led to the emergence of open hole multi-stage fracturing systems. These systems effectively place fractures in specific places in the wellbore, thus increasing the cumulative production in a shorter time frame. Certain reservoirs such as the Bakken, Barnett Shale, Montney and Haynesville Shale cannot be produced using conventional methods. These formations have begun using high tech completion systems capable of mechanically fracturing at certain intervals. An alternative to the plug and perf method, multi-stage fracturing systems are capable of stimulating several stages in a single day. Compared to the weeks required by the plug and perf method, cost-effective multi-stage completion systems are quickly becoming sought after technology by oil and natural gas companies.

### Terminology

***Fracture Gradient :*** The pressure to fracture the formation at a particular depth divided by the depth. A fracture gradient of 18 kPa/m (0.8 psi/foot) implies that at a depth of 3 km (10,000 feet) a pressure of 54 MPa (8,000 psi) will extend a hydraulic fracture.

***ISIP - Initial Shut In Pressure :*** The pressure measured immediately after injection stops. The ISIP provides a measure of the pressure in the fracture at the wellbore by removing contributions from fluid friction.

***Leakoff :*** Loss of fracturing fluid from the fracture channel into the surrounding permeable rock.

***Fracturing fluid :*** The fluid used during a hydraulic fracture treatment of oil, gas or water wells. The fracturing fluid has two major functions:

1. Open and extend the fracture
2. Transport the proppant along the fracture length.

***Proppant :*** Suspended particles in the fracturing fluid that are used to hold fractures open after a hydraulic fracturing treatment, thus producing a conductive pathway that fluids can easily flow along. Naturally occurring sand grains or artificial ceramic material are common proppants used.

***Concise slang :*** "Fracing" (sometimes spelled "fracking" primarily in media) is a shortened version of fracturing.

### Regulation

Hydraulic fracturing for the purpose of oil, natural gas, and geothermal production was exempted under the Safe Drinking Water Act. The decision for this may be due in part to the fact that underground sources of oil and gas already contain high levels of hydrocarbons, including carcinogens like benzene. Hydraulic fracturing for these purposes therefore falls under state jurisdiction. State regulations are inconsistent. All other hydraulic fracturing falls under the EPA regulations as do the handling and disposal of recovered fluids.

About 750 chemicals have been listed as additives for hydraulic fracturing in a report to the US Congress in 2011. Third-party laboratories are performing analysis on soil, air, and water near the fracturing sites to measure the level of contamination by each of the chemicals. Each state has a contact person in charge of such regulation. A map of these contact people can be found at FracFocus.org as well.

### *Environmental Concerns*

A 2004 study by the Environmental Protection Agency (EPA) concluded that the injection of hydraulic fracturing fluids into CBM wells posed minimal threat to underground drinking water sources. This study has been criticised for only focusing on the injection of fracking fluids, while ignoring other aspects of the process such as disposal of fluids, and environmental concerns such as water quality, fish kills and acid burns; the study was also concluded before public complaints of contamination started emerging. Largely on the basis of this study, hydraulic fracturing was in 2005 exempted by US Congress from any regulation under the Safe Drinking Water Act.

With the explosive growth of natural gas wells in the US, researcher Valerie Brown predicted in 2007 that "public exposure to the many chemicals involved in energy development is expected to increase over the next few years, with uncertain consequences." As development of natural gas wells in the U.S. since the year 2000 has increased, so too have claims by private well owners of water contamination. This has prompted EPA and others to re-visit the topic.

Environmental concerns with hydraulic fracturing include the potential contamination of ground water, risks to air quality, the potential migration of gases and hydraulic fracturing chemicals to the surface, the potential mishandling of waste, and the health effects of these. While the EPA recognizes the potential for contamination of water by hydraulic fracturing, EPA Administrator Lisa Jackson testified in a Senate Hearing Committee "I'm not aware of any proven case where the fracking process itself has affected water...". There are, however, documented incidents of contamination. In 2006 drilling fluids and methane were detected leaking from the ground near a gas well in Clark, Wyoming; 8 million cubic feet of methane were eventually released, and shallow groundwater was found to be contaminated. In the town of Dimock, Pennsylvania, 13 water wells were contaminated with methane (one of them blew up), and the gas company, Cabot Oil & Gas, had to financially compensate residents and construct a pipeline to bring in clean water; the company continued to deny, however, that any "of the issues in Dimock have anything to do with hydraulic fracturing".

A Duke University study published in *Proceedings of the National Academy of Sciences* in 2011 examined methane in groundwater in Pennsylvania and New York states overlying the Marcellus Shale and the Utica Shale. It determined that groundwater tended to contain much higher concentrations of methane near fracking wells, with potential explosion hazard; the methane's isotopic signatures and other geochemical indicators were consistent with it originating in the fracked deep shale formations, rather than any other source. Complaints from a few residents on water quality in a developed natural gas field prompted an EPA groundwater investigation in Wyoming. The EPA reported detections of methane and other chemicals such as phthalates in private water wells.

In Pavillion, Wyoming, the EPA discovered traces of methane and foaming agents in several water wells near a gas rig, though it suggested these chemicals might have come from cleaning products. In DISH, Texas, elevated levels of disulphides, benzene, xylenes and naphthalene have been detected in the air, alongside numerous local complaints of headaches, diarrhoea, nosebleeds, dizziness, muscle spasms and other problems. Epidemiological studies that might confirm or rule out any connection between these complaints and fracking are virtually non-existent. Individuals "smell things that don't make them feel well, but we know nothing about cause-and-effect relationships in these cases." In Garfield County, Colorado, another area with a high concentration of drilling rigs, volatile organic compound emissions increased 30% between 2004 and 2006; during the same period there was a rash of health complaints from local residents. The health effects of VOCs are largely unquantified, so any causal relationship is difficult to ascertain; however, some of these chemicals are suspected carcinogens and neurotoxins. Investigators from the Colorado School of Public Health performed a study in Garfield regarding potential adverse health effects, and concluded that residents near gas wells might suffer chemical exposures, accidents from industry operations, and psychological impacts such as depression, anxiety and stress. This study (the only one of its kind to date) was never published, owing to disagreements between community members and the drilling company over the study's methods. In 2010 the film *Gasland* premiered at the Sundance Film Festival. The filmmaker claims that chemicals including toxins, known carcinogens, and heavy metals polluted the ground water near well sites in Pennsylvania, Wyoming, and Colorado.

A 2011 report by the Massachusetts Institute of Technology addressed groundwater contamination, noting "There has been concern that these fractures can also penetrate shallow freshwater zones and contaminate them with fracturing fluid, but there is no evidence that this is occurring". This study blames known instances of methane contamination on a small number of sub-standard operations, and encourages the use of industry best practices to prevent such events from recurring.

Directed by Congress, the U.S. EPA announced in March 2010 that it will examine claims of water pollution related to hydraulic fracturing.

### Chemicals used in Fracturing Fluid

Chemical additives used in fracturing fluids typically make up less than 1% by weight of the total fluid. They are biocides, surfactants, adjusting viscosity, and emulsifiers. Many are used in household products such as cosmetics, lotions, soaps, detergents, furniture polishes, floor waxes, and paints. Some are also used in food products. A list of the chemicals that have been used was published in a U.S. House of Representatives Report. Some of the chemicals pose no known health hazards, some of the chemicals are known carcinogens, some are toxic, and the health hazards for the rest are unknown. The report does not include the concentration of each chemical used or the amount used.

A 2011 study identified 632 chemicals used in natural gas operations. Only 353 of these are well-described in the scientific literature; and of these, more than 75% could affect skin, eyes, respiratory and gastrointestinal systems; roughly 40-50% could affect the brain and nervous, immune and cardiovascular systems and the kidneys; 37% could affect the endocrine system; and 25% were carcinogens or mutagens. The study indicated possible long-term health effects that might not appear immediately. The study recommended full disclosure of all products used, along with extensive air and water monitoring near natural gas operations; it also recommended that fracking's exemption from regulation under the US Safe Drinking Water Act be rescinded.

### *Earthquakes*

The United States Geological Survey (USGS) reports that "Earthquakes induced by human activity have been documented in a few locations" in the United States, Japan, and Canada; "the cause was injection of fluids into deep wells for waste disposal and secondary recovery of oil, and the use of reservoirs for water supplies." The disposal and injection wells referenced are regulated under the Safe Drinking Water Act and UIC laws and are not wells where hydraulic fracturing is generally performed.

### *Greenhouse Gas Emissions*

The use of natural gas rather than oil or coal is sometimes touted as a way of alleviating global warming: natural gas burns more cleanly, and gas power stations can produce up to 50% less greenhouse gases than coal stations. However, an analysis of the well-to-consumer lifecycle of fracked natural gas concluded that 3.6–7.9% of the methane produced by a well will be leaked into the atmosphere during the well's lifetime. Because methane is such a potent greenhouse gas, this means that over short timescales, shale gas is actually worse than coal or oil. Methane gradually breaks down in the atmosphere, so that over long periods it is not as problematic as carbon dioxide; yet even if shale gas is burnt in efficient gas power stations, its greenhouse-gas footprint is still worse than coal or oil for timescales of less than fifty years.

### *Hydraulic Fracturing by Country*

Hydraulic fracturing has become a contentious environmental and health issue with France banning the practice and a moratorium in place in New South Wales (Australia), Karoo basin (South Africa), Quebec (Canada), and some of the states of the US.

### *Australia*

On 21 February 2011, the ABC's investigative journalism program Four Corners aired a program showing incidents of gas leaks into the water basin and evidence of contamination by hydraulic fracturing in Chinchilla, Queensland as a result of drilling carried out by a

Queensland gas company, QGC. There is currently a moratorium in place on the practice of hydraulic fracturing in the state of New South Wales.

### Canada

Fracking has been in use in Canada at an industrial level since the 1990s. Concerns about fracking began, in late July 2011, when the Government of British Columbia gave Talisman Energy a long-term water licence to draw water from the BC Hydro-owned Williston Lake reservoir, for a twenty year term. Fracking has also received criticism in New Brunswick and Nova Scotia, and the Nova Scotia government is currently reviewing the practice, with recomendations expected in March of 2012. The practice has been temporarily suspended, in Quebec, pending an environmental review. The Canadian Centre for Policy Alternatives has also expressed concern.

### France

Hydraulic fracturing was banned in France in 2011 after public pressure.

### New Zealand

In New Zealand, hydraulic fracturing is part of petroleum exploration and extraction on a small scale mainly in Taranaki and concerns have been raised by environmentalists.

### South Africa

There is currently a moratorium on hydraulic fracturing in South Africa's Karoo region despite the interest of several energy companies.

### United Kingdom

Some hydraulic fracturing is currently proceeding in the United Kingdom operated by Cuadrilla and a number of other companies. In Lancashire, operations were suspended after two small earthquakes subsequent to drilling operations. Several protest groups have started to oppose hydraulic fracturing in the UK such as nationwide groups like Frack Off and local groups like Ribble Estuary Against Fracking and The Vale Says No.

## Hydraulic Fracturing in the United States

Considerable controversy surrounds the current implementation of hydraulic fracturing technology in the United States. Hydraulic fracturing, or *fracking*, is the process of utilizing pressurized water, or some other liquid, to fracture rock layers and release petroleum, natural gas, or other substances so that they can be extracted. Environmental safety and health concerns about this process have emerged and are being debated at the state and national levels.

### *History*

Hydraulic fracturing for stimulation of oil and natural gas wells was first used in the United States in 1947. It was first used commercially by Halliburton in 1949, and because of its success in increasing production from oil wells was quickly adopted, and is now used worldwide in tens of thousands of oil and natural gas wells annually. The first industrial use of hydraulic fracturing was as early as 1903, according to T.L. Watson. Before that date, hydraulic fracturing was used at Mt. Airy Quarry, near Mt Airy, North Carolina where it was (and still is) used to separate granite blocks from bedrock.

Volcanic dikes and sills are examples of natural hydraulic fractures. Hydraulic fracturing incorporates results from the disciplines of fracture mechanics, fluid mechanics, solid mechanics, and porous medium flow.

### *Contemporary Use*

Hydraulic fracturing is used in mining operations and to increase the production of oil and natural gas wells. It is also used in remediating subsurface contaminants.

## Environmental and Health Effects

### *Water and Health*

In April 2010 the state of Pennsylvania banned Cabot Oil & Gas Corp. from further drilling in the entire state until it plugs wells believed to be the source of contamination of the drinking water of 14 homes in Dimock Township, Pennsylvania. The investigation was initiated after a water well exploded on New Year's Day in 2009. The state investigation revealed that Cabot Oil & Gas Company "had allowed combustible gas to escape into the region's groundwater supplies."

In the United States, a 2004 Environmental Protection Agency (EPA) study concluded that the process was safe and didn't warrant further study, because there was "no unequivocal evidence" of health risks, and the fluids were neither necessarily hazardous nor able to travel far underground. That study, however, was not intended as a general study of hydraulic fracturing, but only of its use in coalbed methane deposits, and the study did not consider impacts above ground. The EPA report did find uncertainties in knowledge of how fracturing fluid migrates through rocks, and upon its release service companies voluntarily agreed to stop using diesel fuel as a component of fracturing fluid in coalbed methane walls due to public concerns of its potential as a source of benzene contamination. Environmental group Riverkeeper presented a report to the EPA of over 100 cases of contamination. It has published a report of various environmental impacts using reports from federal and state regulators.

The increased use of hydraulic fracturing has prompted more speculation about its environmental dangers. A 2008 investigation of benzene contamination in Colorado and Wyoming led some EPA officials to suggest hydraulic fracturing as a culprit. One of the authors of the 2004 EPA report states that it has been misconstrued by the gas-drilling industry.

A 2011 study by Congressional Democrats found that, in the process of hydraulic fracturing, "oil and gas companies injected hundreds of millions of gallons of hazardous or carcinogenic chemicals into wells in more than 13 states from 2005 to 2009," according to the New York Times. A 2011 investigation by the New York Times based on various leaked EPA documents found that hydraulic fracturing had resulted in significant increases of radioactive material including radium and carcinogens including benzene in major rivers and watersheds. At one site the amount of benzene discharged into the Allegheny River after treatment was 28 times accepted levels for drinking water.

A 2011 peer-reviewed study found, on average, methane concentrations 17 times above normal in samples taken from water wells near shale gas drilling sites employing hydraulic fracturing. Water samples from 68 private water wells in the states of Pennsylvania and New York were tested and some were found to have extremely high concentrations of methane: 64 milligrams of methane per liter of drinking water, compared with a normal level of one milligram or lower. According to one of the authors of the study, "That sort of concentration is up at a level where people worry about an explosion hazard." The average concentration of methane in the water wells near drilling sites lies within a range that, according to the U.S. Department of the Interior, is dangerous and requires urgent "hazard mitigation" action. The research was conducted by scientists at Duke University and what they found was that "levels of flammable methane gas in drinking water wells increased to dangerous levels when those water supplies were close to natural gas wells. They also found that the type of gas detected at high levels in the water was the same type of gas that energy companies were extracting from thousands of feet underground, strongly implying that the gas may be seeping underground through natural or manmade faults and fractures, or coming from cracks in the well structure itself ".

Methane contamination has been a common complaint among people who live near natural gas drilling areas. In 2009, a Propublica investigation revealed that methane contamination is widespread, "methane related to the natural gas industry has contaminated water wells in at least seven Pennsylvania counties since 2004". Because of this contamination, several homes have blown up after gas seeped into their water supplies; there have been reports of house explosions in Pennsylvania and Ohio. In one case in 2004, a methane leak caused an explosion that killed a couple and their 17 month old grandson.

A study published in May 2011 concluded that fracking has seriously contaminated shallow groundwater supplies in northeast Pennsylvania with flammable methane. However the study does not discuss how pervasive such contamination might be in other areas where drilling for shale gas has taken place.

### *Impact on Drinking Water Supplies*

As also noted in the general article on Hydraulic Fracturing, there are also concerns about hydraulic fracturing's impact on drinking water supplies. The procedure uses millions of gallons of fresh water per well. Some states, such as Texas, Florida, Arizona, and California are trying to get ahead of drinking water shortages by recycling sewage (also called effluent, gray, and reclaimed water). This form of processing cannot, however, remove many of the toxins used in hydraulic fracturing or all pathogens present in wastewater. Its effectiveness and safety in converting wastewater into drinking water is being debated. This technology has been used in the US space program for years.

### *Well Blowouts and Spills of Fracturing Fluids*

A well blowout in Clearfield County, Pennsylvania on June 3, 2010, sent more than 35,000 gallons of hydraulic fracturing fluids into the air and onto the surrounding landscape in a forested area. Campers were evacuated and the company EOG Resources and the well completion company C.C. Forbes have been ordered to cease all operations in the state of Pennsylvania pending investigation. The Pennsylvania Department of Environmental Protection has called this a "serious incident".

### Natural Gas Drilling and Seismic Events

Injection of fluid into subsurface geological structures, such as faults and fractures, reduces the effective normal stress acting across these structures.

If sufficient shear stress is present, the structure may slip in shear and generate seismic events over a range of magnitudes; it is possible that natural gas drilling may have caused earthquakes in North Texas. Subsidence is not directly caused by hydraulic fracturing but may occur after considerable production of oil or ground water. Subsidence occurs over reservoirs whether they have been subject to hydraulic fracturing or not because it is a result of producing fluids from the reservoir and lowering the reservoir pore pressure. The subsidence process can be associated with some seismicity. Reports of minor tremors of no greater than 2.8 on the Richter scale were reported on June 2, 2009 in Cleburne, Texas, the first in the town's 140-year history.

Fracking connected to deep well waste water disposal has been banned in Arkansas due to the risk of contaminating aquifers if the earthquakes it induced continued.

### Air and Health

A potential hazard that is commonly overlooked is the venting of bulk sand silos directly to atmosphere. When they are being filled, or emptied during the fracture, a fine cloud of silica particulate will be vented directly into atmosphere. This dust has the potential to travel many kilometres on the wind directly into populated areas. While the immediate personnel are wearing personal protective equipment, other people in the area of a well fracture can potentially be exposed. Many particulates and chemicals can be released into the atmosphere, such as sulfuric Oxide, nitrous oxides, volatile organic compounds (VOC), benzene, toluene, diesel fuel, hydrogen sulfide which can have serious health implications.

### *Other Consequences*

A 2011 Cornell University study found that natural gas extracted by hydraulic fracturing may contribute as much to global warming as coal, or more so. The authors of the study acknowledge, however, that the data they used were not the best available, so this has left the study open to interpretation. The natural gas industry has noted significant problems with the study's methodology, including comments from the authors that acknowledge the data may be unreliable.

There are also potential community complications as a result of fracking. When drilling companies move into a new area the population increases and with it comes problems related to population boom. There is the potential for noise and light pollution complaints, reports of crime can go up, motor vehicle accidents increase, traffic and road degradation increase, sexually transmitted infections increase, and strain on schools are all some potential problems facing communities where gas drilling is nearby.

In Garfield county, Colorado the Colorado School of Public Health released a second draft report released the Battlement Mesa Health Impact Assessment on March 1, 2011 for public comment.

University of Michigan-Ann Arbor in May 2011 began studying the effects of natural gas fracking on the Great Lakes.

Congress has requested that the EPA undertake a new, broader study of hydraulic fracturing. The report is due to be released in 2012.

### *Occupational Hazards*

The EPA states in their Hydraulic Fracturing Study Plan (2011) that the exposure to hydraulic fracturing chemicals in an occupational setting needs to be examined to determine the acute and chronic effects on health. The exposure risks such as "transport, mixing, delivery, and potential accidents" have not been properly assessed.

Hydraulic fracturing also affects individuals in close proximity to the activity, like the case previously discussed about the nurse who became ill after exposure from treating a hydraulic fracturing worker (Frankowski, 2008).

### *Economic Impacts*

The economic impacts of hydraulic fracturing include property owners that receive funds, an increase in jobs, and an increase in business. The EPA states that it is unclear on a local level how and for how long hydraulic fracturing affects a community economically. It is hypothesized that hydraulic fracturing may not provide jobs to local communities due to the specialized nature of hydraulic fracturing tasks. Also, communities' local resources could potentially be taxed due to the increase in industry traffic or if an accident occurs.

Furthermore, a June, 2011 New York Times investigation of industrial emails and internal documents found that the financial benefits of unconventional shale gas extraction may be less than previously thought, due to companies intentionally overstating the productivity of their wells and the size of their reserves.

### Lawsuits

In September 2010, a lawsuit was filed in Pennsylvania alleging that Southwestern Energy Company contaminated aquifers through a defective cement casing in the well.

In May 2011, Jessica Ernst launched a multi million dollar lawsuit against Encana Corporation, the Alberta Energy Resources Conservation Board, and the Alberta government for contamination of her property and drinking water due to Encana's fracking program. Encana fractures rock to extract coal bed methane, much as fracking is used to extract natural gas from shale. (In March, after a public hearing, Quebec put a moratorium on shale gas exploration pending a full environmental assessment of the potential damage from fracking.) According to the Statement of Claim, many Albertans depend on drinking water from coal bed aquifers, but Ms.

Ernst's water is now so contaminated that it can be lit on fire. She is also suing Alberta's oil and gas regulator, alleging that it not only tolerated illegal behaviour by Encana and failed to protect her, but actively attempted to silence her complaints, and that Alberta Environment showed bad faith in "investigating" those complaints. The lawsuit, together with the Quebec moratorium, signals the likelihood of stronger environmental regulations of fracking in the pursuit of shale gas or coalbed methane.

In June 2011, Northeast Natural Energy sued the town of Morgantown, West Virginia, for its ban on hydraulic fracturing in the Marcellus Shale within a mile of the town's borders. The ban had been initiated to protect the municipal water supply as well as the town's inhabitants, in the absence at the state level of regulations specific to hydraulic fracturing.

### *Regulation*

On August 7, 1997, the Eleventh Circuit Court ordered the United States Environmental Protection Agency to reevaluate its stance on hydraulic fracturing based on a lawsuit brought by the Legal Environmental Assistance Foundation. Up until that decision, the EPA deemed that hydraulic fracturing did not fall under the rules in the Safe Drinking Water Act. While the impact of this decision was localized to Alabama, it forced the EPA to evaluate its oversight responsibility under the Safe Drinking Water Act for hydraulic fracturing. In 2004, the EPA released a study that concluded the threat to drinking water from hydraulic fracturing was "minimal" and the Energy Policy Act of 2005 exempted fractured wells from being re-classified as injection wells, which would place them under federal regulation under the Safe Drinking Water Act. which was originally intended to regulate disposal wells.

In April 2011, the Ground Water Protection Council, in conjunction with the industry, began releasing well-by-well lists of hydraulic fracturing chemicals at.

A complete listing of the specific chemical formulation of additives used in hydraulic fracturing operations is not currently made available to landowners, neighbors, local officials, or health care providers. This practice is under scrutiny as well.

Two studies released in 2009, one by the U.S. Department of Energy and the other released by the Ground Water Protection Council, address hydraulic fracturing safety concerns. Chemicals which can be used in the fracturing fluid include kerosene, benzene, toluene, xylene, and formaldehyde. These chemicals are not directly used as treating chemical additives but can be a small component of the specific chemicals used in the job.

On June 8, 2010 the Wyoming Oil and Gas Conservation Commission voted to require full disclosure of the hydraulic fracturing fluids used in natural gas exploration. This may aid in tracking pollutants that have migrated from hydraulically fractured gas wells.

Congress has been urged to repeal the 2005 regulatory exemption under the Energy Policy Act of 2005. The FRAC Act, introduced in June 2009, would eliminate the exemption and might allow producing wells to be reclassified as injection wells placing them under federal jurisdiction in states without approved UIC programs.

The New York City watershed includes a large area of the Marcellus shale formation. The NYC Dept. of Environmental Protection's position: "While DEP is mindful of the potential economic opportunity that this represents for the State, hydraulic fracturing poses an unacceptable threat to the unfiltered water supply of nine million New Yorkers and cannot safely be permitted with the New York City watershed."

The New York State assembly voted 93 to 43, Nov. 30, 2010, to place a moratorium or freeze on hydraulic fracturing to give the state more time to undertake safety and environmental concerns.

At the municipal level, some towns and cities in central New York state have moved to regulate drilling by hydraulic fracturing and its attendant effects, either by banning it within municipal limits, maintaining the option to do so in the future, or banning wastewater from the drilling process from municipal water treatment plants.

### *EPA Hydraulic Fracturing Study*

The purpose of the EPA study regarding Hydraulic Fracturing is to examine the effects of hydraulic fracturing on the water supply, specifically for human consumption. The research aims to examine the full scope of the water pathway as it moves through the hydraulic fracturing process, including water that is used for the construction of the wells, the fracturing mixture, and subsequent removal and disposal. Fundamental research questions include:

- How might large volume water withdrawals from ground and surface water impact drinking water resources?
- What are the possible impacts of releases of flowback and produced water on drinking water resources?
- What are the possible impacts of inadequate treatment of hydraulic fracturing wastewaters on drinking water resources?

The Scientific Advisory Board reviewed the study plan in early March 2011. In June 2011, the EPA announced the locations of its five retrospective case studies, which will examine existing hydrofracking sites for evidence of drinking water contamination. They are:

- Bakken Shale - Kildeer, and Dunn Counties, North Dakota
- Barnett Shale - Wise and Denton Counties, Texas
- Marcellus Shale - Bradford and Susquehanna Counties, Pennsylvania

- Marcellus Shale - Washington County, Pennsylvania
- Raton Basin - Las Animas County, Colorado.

Research should be completed by the end of 2012, and the EPA's Hydraulic Fracturing Report is expected to be completed in 2014.

### *Fracturing Responsibility and Awareness of Chemicals Act*

In June 2009 two identical bills named the Fracturing Responsibility and Awareness of Chemicals Act were introduced to both the United States House and the Senate. The House bill was introduced by representatives Diana DeGette, D-Colo., Maurice Hinchey D-N.Y., and Jared Polis, D-Colo. The Senate version was introduced by senators Bob Casey, D-Pa., and Chuck Schumer, D-N.Y. These bills are designed to amend the Safe Drinking Water Act. This would allow the Environmental Protection Agency to regulate hydraulic fracturing that occurs in states which have not taken primacy in UIC regulation. The bill required the energy industry to reveal what chemicals are being used in the sand-water mixture. The 111th Congress adjourned (Jan. 3, 2011) without taking action on the FRAC Act. The 112th Congress has not re-introduced the bill or an equivalent.

### *Halliburton's Hydraulic Fracturing Operations*

As mandated by EPA subpoena, Halliburton has released hydraulic fracturing operations. Details regarding Halliburton's fracturing process and fracturing chemicals can be found on their website.

### *Fracturing Method: High Pressure by Combustible Gas Mixtures or Driving Forces*

Fracturing is done by pumping in liquids at high pressure. In the past, combustible gas mixtures, driving liquids or explosives generated high-pressure high-speed gas flow (TNT or PETN up to 1,900,000 psi). In the late 1960s and early 1970s, as part of Operation Plowshare, underground nuclear explosions were tested for natural gas stimulation. The Rulison explosion multiplied the accessibility of the gas, but the gas was contaminated and unmarketable. Stimulation by nuclear explosion is not currently utilized by the industry.

## List of Additives for Hydraulic Fracturing

About 750 chemicals have been listed as additives for hydraulic fracturing in a report to the US Congress in 2011. The following is a partial list of the chemical constituents in additives used in fracturing operations, as indicated by the New York State Department of Environmental Conservation. Each state has a contact person in charge of the fracking regulations of these additives.

| *CAS Number* | *Chemical Constituent* |
|---|---|
| 2634-33-5 | 1,2-Benzisothiazolin-2-one/1,2-benzisothiazolin-3-one |
| 95-63-6 | 1,2,4-trimethylbenzene |
| 123-91-1 | 1,4-Dioxane |
| 3452-07-1 | 1-eicosene |
| 629-73-2 | 1-hexadecene |
| 112-88-9 | 1-octadecene |
| 1120-36-1 | 1-tetradecene |
| 10222-01-2 | 2,2 Dibromo-3-nitrilopropionamide, a biocide |
| 27776-21-2 | 2,2'-azobis-{2-(imidazlin-2-yl)propane}-dihydrochloride |
| 73003-80-2 | 2,2-Dibromomalonamide |
| 15214-89-8 | 2-Acrylamido-2-methylpropane sulphonic acid sodium salt polymer |
| 46830-22-2 | 2-acryloyloxyethyl(benzyl)dimethylammonium chloride |
| 52-51-7 | 2-Bromo-2-nitro-1,3-propanediol |
| 111-76-2 | 2-Butoxy ethanol |
| 1113-55-9 | 2-Dibromo-3-Nitriloprionamide (2-Monobromo-3-nitriilopropionamide) |
| 104-76-7 | 2-Ethyl Hexanol |
| 67-63-0 | 2-Propanol/Isopropyl Alcohol/Isopropanol/Propan-2-ol |
| 26062-79-3 | 2-Propen-1-aminium, N,N-dimethyl-N-2-propenyl-chloride, homopolymer |
| 9003-03-6 | 2-propenoic acid, homopolymer, ammonium salt |
| 25987-30-8 | 2-Propenoic acid, polymer with 2 p-propenamide, sodium salt/Copolymer of acrylamide and sodium acrylate |
| 71050-62-9 | 2-Propenoic acid, polymer with sodium phosphinate (1:1) |
| 66019-18-9 | 2-propenoic acid, telomer with sodium hydrogen sulfite |
| 107-19-7 | 2-Propyn-1-ol/Propargyl alcohol |
| 51229-78-8 | 3,5,7-Triaza-1-azoniatricyclo[3.3.1.13,7]decane, 1-(3-chloro-2-propenyl)-chloride, |
| 115-19-5 | 3-methyl-1-butyn-3-ol |
| 127087-87-0 | 4-Nonylphenol Polyethylene Glycol Ether Branched/ Nonylphenol ethoxylated/Oxyalkylated Phenol |
| 64-19-7 | Acetic acid |
| 68442-62-6 | Acetic acid, hydroxy-, reaction products with triethanolamine |

| | |
|---|---|
| 108-24-7 | Acetic Anhydride |
| 67-64-1 | Acetone |
| 79-06-1 | Acrylamide |
| 38193-60-1 | Acrylamide - sodium 2-acrylamido-2-methylpropane sulfonate copolymer |
| 25085-02-3 | Acrylamide - Sodium Acrylate Copolymer or Anionic Polyacrylamide |
| 69418-26-4 | Acrylamide polymer with N,N,N-trimethyl-2[1-oxo-2-propenyl]oxy Ethanaminium chloride |
| 15085 | -02-3 Acrylamide-sodium acrylate copolymer |
| 68551-12-2 | Alcohols, C12-C16, Ethoxylated (a.k.a. Ethoxylated alcohol) |
| 64742-47-8 | Aliphatic Hydrocarbon/Hydrotreated light distillate/ Petroleum Distillates/Isoparaffinic Solvent/Paraffin Solvent/Napthenic Solvent |
| 64743-02-8 | Alkenes |
| 68439-57-6 | Alkyl (C14-C16) olefin sulfonate, sodium salt |
| 9016-45-9 | Alkylphenol ethoxylate surfactants |
| 1327-41-9 | Aluminum chloride |
| 73138-27-9 | Amines, C12-14-tert-alkyl, ethoxylated |
| 71011-04-6 | Amines, Ditallow alkyl, ethoxylated |
| 68551-33-7 | Amines, tallow alkyl, ethoxylated, acetates |
| 1336-21-6 | Ammonia |
| 631-61-8 | Ammonium acetate |
| 68037-05-8 | Ammonium Alcohol Ether Sulfate |
| 7783-20-2 | Ammonium bisulfate |
| 10192-30-0 | Ammonium bisulfite |
| 12125-02-9 | Ammonium chloride |
| 7632-50-0 | Ammonium citrate |
| 37475-88-0 | Ammonium Cumene Sulfonate |
| 1341-49-7 | Ammonium hydrogen-difluoride |
| 6484-52-2 | Ammonium nitrate |
| 7727-54-0 | Ammonium Persulfate/Diammonium peroxidisulphate |
| 1762-95-4 | Ammonium Thiocyanate |
| 7664-41-7 | Aqueous ammonia |
| 121888-68-4 | Bentonite, benzyl(hydrogenated tallow alkyl) dimethylammonium stearate complex/organophilic clay |
| 71-43-2 | Benzene |
| 119345-04-9 | Benzene, 1,1'-oxybis, tetratpropylene derivatives, sulfonated, sodium salts |

| | |
|---|---|
| 74153-51-8 | Benzenemethanaminium, N,N-dimethyl-N-[2-[(1-oxo-2-propenyl)oxy]ethyl]-, chloride, polymer with 2-propenamide |
| 10043-35-3 | Boric acid |
| 1303-86-2 | Boric oxide/Boric Anhydride |
| 71-36-3 | Butan-1-ol |
| 68002-97-1 | C10 - C16 Ethoxylated Alcohol |
| 68131-39-5 | C12-15 Alcohol, Ethoxylated |
| 10043-52-4 | Calcium chloride |
| 124-38-9 | Carbon dioxide |
| 68130-15-4 | Carboxymethylhydroxypropyl guar |
| 9012-54-8 | Cellulase/Hemicellulase Enzyme |
| 9004-34-6 | Cellulose |
| 10049-04-4 | Chlorine dioxide |
| 77-92-9 | Citric Acid |
| 94266-47-4 | Citrus Terpenes |
| 61789-40-0 | Cocamidopropyl betaine |
| 68155-09-9 | Cocamidopropylamine Oxide |
| 68424-94-2 | Coco-betaine |
| 7758-98-7 | Copper(II) sulfate |
| 31726-34-8 | Crissanol A-55 |
| 14808-60-7 | Crystalline Silica (Quartz) |
| 7447-39-4 | Cupric chloride dihydrate |
| 1120-24-7 | Decyldimethyl Amine |
| 2605-79-0 | Decyl-dimethyl Amine Oxide |
| 3252-43-5 | Dibromoacetonitrile |
| 25340-17-4 | Diethylbenzene |
| 111-46-6 | Diethylene glycol |
| 22042-96-2 | Diethylenetriamine penta (methylenephonic acid) sodium salt |
| 28757-00-8 | Diisopropyl naphthalenesulfonic acid |
| 68607-28-3 | Dimethylcocoamine, bis(chloroethyl) ether, diquaternary ammonium salt |
| 7398-69-8 | Dimethyldiallylammonium chloride |
| 25265-71-8 | Dipropylene glycol |
| 139-33-3 | Disodium Ethylene Diamine Tetra Acetate |
| 5989-27-5 | D-Limonene |
| 123-01-3 | Dodecylbenzene |
| 27176-87-0 | Dodecylbenzene sulfonic acid |

| | |
|---|---|
| 42504-46-1 | Dodecylbenzenesulfonate isopropanolamine |
| 50-70-4 | D-Sorbitol/Sorbitol |
| 37288-54-3 | Endo-1,4-beta-mannanase, or Hemicellulase |
| 149879-98-1 | Erucic Amidopropyl Dimethyl Betaine |
| 89-65-6 | Erythorbic acid, anhydrous |
| 54076-97-0 | Ethanaminium, N,N,N-trimethyl-2-[(1-oxo-2-propenyl)oxy]-, chloride, homopolymer |
| 107-21 | -1 Ethane-1,2-diol/Ethylene Glycol |
| 9002-93-1 | Ethoxylated 4-tert-octylphenol |
| 68439-50-9 | Ethoxylated alcohol |
| 126950-60-5 | Ethoxylated alcohol |
| 67254-71-1 | Ethoxylated alcohol (C10-12) |
| 68951-67-7 | Ethoxylated alcohol (C14-15) |
| 68439-46-3 | Ethoxylated alcohol (C9-11) |
| 66455-15-0 | Ethoxylated Alcohols |
| 84133-50-6 | Ethoxylated Alcohols (C12-14 Secondary) |
| 68439-51-0 | Ethoxylated Alcohols (C12-14) |
| 78330-21-9 | Ethoxylated branch alcohol |
| 34398-01-1 | Ethoxylated C11 alcohol |
| 61791-12-6 | Ethoxylated Castor Oil |
| 61791-29-5 | Ethoxylated fatty acid, coco |
| 61791-08-0 | Ethoxylated fatty acid, coco, reaction product with ethanolamine |
| 68439-45-2 | Ethoxylated hexanol |
| 9036-19-5 | Ethoxylated octylphenol |
| 9005-67-8 | Ethoxylated Sorbitan Monostearate |
| 9004-70-3 | Ethoxylated Sorbitan Trioleate |
| 64-17-5 | Ethyl alcohol/ethanol |
| 100-41-4 | Ethyl Benzene |
| 97-64-3 | Ethyl lactate |
| 9003-11-6 | Ethylene Glycol-Propylene Glycol Copolymer (Oxirane, methyl-, polymer with oxirane) |
| 75-21-8 | Ethylene oxide |
| 5877-42-9 | Ethyloctynol |
| 68526-86-3 | Exxal 13 |
| 61790-12-3 | Fatty Acids |
| 68188-40-9 | Fatty acids, tall oil reaction products w/acetophenone, formaldehyde & thiourea |
| 9043-30-5 | Fatty alcohol polyglycol ether surfactant |

| | |
|---|---|
| 7705-08-0 | Ferric chloride |
| 7782-63-0 | Ferrous sulfate, heptahydrate |
| 50-00-0 | Formaldehyde |
| 29316-47-0 | Formaldehyde polymer with 4,1,1-dimethylethyl phenolmethyl oxirane |
| 153795-76-7 | Formaldehyde, polymers with branched 4-nonylphenol, ethylene oxide and propylene oxide |
| 75-12-7 | Formamide |
| 64-18-6 | Formic acid |
| 110-17-8 | Fumaric acid |
| 65997-17-3 | Glassy calcium magnesium phosphate |
| 111-30-8 | Glutaraldehyde |
| 56-81-5 | Glycerol/glycerine |
| 9000-30-0 | Guar Gum |
| 64742-94-5 | Heavy aromatic petroleum naphtha |
| 9025-56-3 | Hemicellulase |
| 7647-01-0 | Hydrochloric Acid/Hydrogen Chloride/muriatic acid |
| 7722-84-1 | Hydrogen peroxide |
| 79-14-1 | Hydroxy acetic acid |
| 35249-89-9 | Hydroxyacetic acid ammonium salt |
| 9004-62-0 | Hydroxyethyl cellulose |
| 5470-11-1 | Hydroxylamine hydrochloride |
| 39421-75-5 | Hydroxypropyl guar |
| 35674-56-7 | Isomeric Aromatic Ammonium Salt |
| 64742-88-7 | Isoparaffinic Petroleum Hydrocarbons, Synthetic |
| 64-63-0 | Isopropanol |
| 98-82-8 | Isopropylbenzene (cumene) |
| 68909-80-8 | Isoquinoline, reaction products with benzyl chloride and quinoline |
| 8008-20-6 | Kerosene |
| 64742-81-0 | Kerosine, hydrodesulfurized |
| 63-42-3 | Lactose |
| 64742-95-6 | Light aromatic solvent naphtha |
| 1120-21-4 | Light Paraffin Oil |
| 14807-96-6 | Magnesium Silicate Hydrate (Talc) |
| 1184-78-7 | methanamine, N,N-dimethyl-, N-oxide |
| 67-56-1 | Methanol |
| 68891-11-2 | Methyloxirane polymer with oxirane, mono (nonylphenol) ether, branched |

| | |
|---|---|
| 8052-41-3 | Mineral spirits/Stoddard Solvent |
| 141-43-5 | Monoethanolamine |
| 44992-01-0 | N,N,N-trimethyl-2[1-oxo-2-propenyl]oxy Ethanaminium chloride |
| 64742-48-9 | Naphtha (petroleum), hydrotreated heavy |
| 91-20-3 | Naphthalene |
| 38640-62-9 | Naphthalene bis(1-methylethyl) |
| 93-18-5 | Naphthalene, 2-ethoxy- |
| 68909-18-2 | N-benzyl-alkyl-pyridinium chloride |
| 68139-30-0 | N-Cocoamidopropyl-N,N-dimethyl-N-2-hydroxypropylsulfobetaine |
| 7727-37-9 | Nitrogen, Liquid form |
| 68412-54-4 | Nonylphenol Polyethoxylate |
| 121888-66-2 | Organophilic Clays |
| 64742-65-0 | Petroleum Base Oil |
| 64741-68-0 | Petroleum naphtha |
| 70714-66-8 | Phosphonic acid, [[(phosphonomethyl)imino]bis[2,1-ethanediylnitrilobis(methylene)]]tetrakis-, ammonium salt |
| 8000-41-7 | Pine Oil |
| 60828-78-6 | Poly(oxy-1,2-ethanediyl), a-[3,5-dimethyl-1-(2-methylpropyl)hexyl]-w-hydroxy- |
| 25322-68-3 | Poly(oxy-1,2-ethanediyl), a-hydro-w-hydroxy/Polyethylene Glycol |
| 24938-91-8 | Poly(oxy-1,2-ethanediyl), α-tridecyl-ù-hydroxy- |
| 51838-31-4 | Polyepichlorohydrin, trimethylamine quaternized |
| 56449-46-8 | Polyethlene glycol oleate ester |
| 62649-23-4 | Polymer with 2-propenoic acid and sodium 2-propenoate |
| 9005-65-6 | Polyoxyethylene Sorbitan Monooleate |
| 61791-26-2 | Polyoxylated fatty amine salt |
| 127-08-2 | Potassium acetate |
| 12712-38-8 | Potassium borate |
| 1332-77-0 | Potassium borate |
| 20786-60-1 | Potassium Borate |
| 584-08-7 | Potassium carbonate |
| 7447-40-7 | Potassium chloride |
| 590-29-4 | Potassium formate |
| 1310-58-3 | Potassium Hydroxide |
| 13709-94-9 | Potassium metaborate |
| 24634-61-5 | Potassium sorbate |

| | |
|---|---|
| 112926-00-8 | Precipitated silica/silica gel |
| 57-55-6 | Propane-1,2-diol, or Propylene glycol |
| 107-98-2 | Propylene glycol monomethyl ether |
| 68953-58-2 | Quaternary Ammonium Compounds |
| 62763-89-7 | Quinoline,2-methyl-, hydrochloride |
| 15619-48-4 | Quinolinium, 1-(phenylmethl),chloride |
| 7631-86-9 | Silica, Dissolved |
| 5324-84-5 | Sodium 1-octanesulfonate |
| 127-09-3 | Sodium acetate |
| 95371-16-7 | Sodium Alpha-olefin Sulfonate |
| 532-32-1 | Sodium benzoate |
| 144-55-8 | Sodium bicarbonate |
| 7631-90-5 | Sodium bisulfate |
| 7647-15-6 | Sodium bromide |
| 497-19-8 | Sodium carbonate |
| 7647-14-5 | Sodium Chloride |
| 7758-19-2 | Sodium chlorite |
| 3926-62-3 | Sodium chloroacetate |
| 68-04-2 | Sodium citrate |
| 6381-77-7 | Sodium erythorbate/isoascorbic acid, sodium salt |
| 2836-32-0 | Sodium Glycolate |
| 1310-73-2 | Sodium Hydroxide |
| 7681-52-9 | Sodium hypochlorite |
| 7775-19-1 | Sodium Metaborate .8H2O |
| 10486-00-7 | Sodium perborate tetrahydrate |
| 7775-27-1 | Sodium persulfate |
| 9003-04-7 | Sodium polyacrylate |
| 7757-82-6 | Sodium sulfate |
| 1303-96-4 | Sodium tetraborate decahydrate |
| 7772-98-7 | Sodium thiosulfate |
| 1338-43-8 | Sorbitan Monooleate |
| 57-50-1 | Sucrose |
| 5329-14-6 | Sulfamic acid |
| 112945-52-5 | Synthetic Amorphous/Pyrogenic Silica/Amorphous Silica |
| 68155-20-4 | Tall Oil Fatty Acid Diethanolamine |
| 8052-48-0 | Tallow fatty acids sodium salt |
| 72480-70-7 | Tar bases, quinoline derivs., benzyl chloride-quaternized |

| | |
|---|---|
| 68647-72-3 | Terpene and terpenoids |
| 68956-56-9 | Terpene hydrocarbon byproducts |
| 533-74-4 | Tetrahydro-3,5-dimethyl-2H-1,3,5-thiadiazine-2-thione (a.k.a. Dazomet) |
| 55566-30-8 | Tetrakis(hydroxymethyl)phosphonium sulfate (THPS) |
| 75-57-0 | Tetramethyl ammonium chloride |
| 64-02-8 | Tetrasodium Ethylenediaminetetraacetate |
| 68-11-1 | Thioglycolic acid |
| 62-56-6 | Thiourea |
| 68527-49-1 | Thiourea, polymer with formaldehyde and 1-phenylethanone |
| 108-88-3 | Toluene |
| 81741-28-8 | Tributyl tetradecyl phosphonium chloride |
| 68299-02-5 | Triethanolamine hydroxyacetate |
| 112-27-6 | Triethylene glycol |
| 52624-57-4 | Trimethylolpropane, Ethoxylated, Propoxylated |
| 150-38-9 | Trisodium Ethylenediaminetetraacetate |
| 5064-31-3 | Trisodium Nitrilotriacetate |
| 7601-54-9 | Trisodium orthophosphate |
| 57-13-6 | Urea |
| 25038-72-6 | Vinylidene Chloride/Methylacrylate Copolymer |
| 7732-18-5 | Water |
| 1330-20-7 | Xylene |
| | Aliphatic acids |
| | Aliphatic alcohol glycol ether |
| | Alkyl Aryl Polyethoxy Ethanol |
| | Alkylaryl Sulfonate |
| | Aromatic hydrocarbons |
| | Aromatic ketones |
| | Oxyalkylated alkylphenol |
| | Petroleum distillate blend |
| | Polyethoxylated alkanol |
| | Polymeric Hydrocarbons |
| | Salt of amine-carbonyl condensate |
| | Salt of fatty acid/polyamine reaction product |
| | Sugar |
| | Surfactant blend |

## Shale Oil Extraction

Shale oil extraction is an industrial process for unconventional oil production. This process converts kerogen in oil shale into shale oil by pyrolysis, hydrogenation, or thermal dissolution. The resultant shale oil is used as fuel oil or upgraded to meet refinery feedstock specifications by adding hydrogen and removing sulfur and nitrogen impurities.

Shale oil extraction is usually performed above ground (*ex situ* processing) by mining the oil shale and then treating it in processing facilities. Other modern technologies perform the processing underground (on-site or *in situ* processing) by applying heat and extracting the oil via oil wells.

The earliest description of the process dates to the 10th century. In 1684, Great Britain granted the first formal extraction process patent. Extraction industries and innovations became widespread during the 19th century. The industry shrank in the mid-20th century following the discovery of large reserves of conventional oil, but high petroleum prices at the beginning of the 21st century have led to renewed interest, accompanied by the development and testing of newer technologies.

As of 2010, major long-standing extraction industries are operating in Estonia, Brazil, and China. Its economic viability usually requires a lack of locally available crude oil. National energy security issues have also played a role in its development. Critics of shale oil extraction pose questions about environmental management issues, such as waste disposal, extensive water use, waste water management, and air pollution.

## History of the Oil Shale Industry

The history of the oil shale industry started in ancient times. The modern industrial use of oil shale for oil extraction dates to the mid-19th century and started growing just before World War I because of the mass production of automobiles and trucks and the supposed shortage of gasoline for transportation needs. Between the World Wars oil shale projects were begun in several countries, but after World War II, the oil shale industry was phased out in almost all countries because of high processing costs and the discovery of large supplies of easily-accessible crude oil. As of 2010, oil shale was commercially used in Estonia, China and Brazil, while several countries are considering to start or re-start commercial use of oil shale.

## Early History

Humans have used oil shale as a fuel since prehistoric times, since it generally burns without any processing. It was also used for decorative purposes and construction. Britons of the Iron Age used to polish and form oil shale into ornaments. Around 3000 BC, "rock oil" was used in Mesopotamia for road construction and making architectural adhesives. As a decorative material, oil shale was also used over the Greek, Roman, Byzantinian, Umayyad and Abbasid periods to decorate mosaics and floors of the palaces, churches and mosques. Shale oil was used for medical and military purposes. Mesopotamians used it for medical purposes and for caulking ships, Mongols used to cap their arrows with flaming oil shale. In 10th century, the Arabian physician Masawaih al-Mardini (Mesue the Younger) described a method of extraction of oil from "some kind of bituminous shale".

In the early 14th century, the first use of shale oil was recorded in Switzerland and Austria. In 1350, a knight Berthold von Eberhausen was awarded a right to exploit the Seefeld oil shale in Tyrol. Oil shale was used for production of shale oil using an early retorting method of heating the crushed oil shale put in crucibles. The healing properties of a mineral oil distilled from oil shale were noted in 1596 by the personal physician of the Duke of Württemberg Frederick I. In Skåne, the Swedish alum shale dating from the Cambrian and Ordovician periods was used for extracting potassium aluminium sulfate by roasting it over fire as early as 1637. In Italy, shale oil was used to light the streets of Modena at the turn of the 17th century. The British Crown granted a patent in 1694 to three persons who had "found a way to extract and make great quantities of pitch, tarr, and oyle out of a sort of stone." Shale oil was produced by extracting Shropshire oil shale. Later sold as Betton's British Oil, the distilled product was said to have been "tried by divers persons in Aches and Pains with much benefit. In 1781, Archibald Cochrane, 9th Earl of Dundonald, registered a patent for an extraction process to produce tar, pitch and oil from coal and bituminous shales, using masonry retorts and wooden condensers.

## Start of the Modern Industry

The modern industrial use of oil shale for oil extraction started in France, where oil shale commercial mining began in Autun in 1837. The shale oil production started in 1838 by using Selligue process, invented by Alexander Selligue. In 1847 the Scottish chemist James

Young prepared "lighting oil," lubricating oil and wax from cannel coal and since 1862 from torbanite. In 1850 he patented the process of cracking oil. Commercial scale shale oil extraction in Scotland started in 1859 by Robert Bell in Broxburn, West Lothian. About the same time Germany began exploiting its deposits. During the second half of the 19th century shale oil extraction industries were initiated also in Sweden, Australia, Brazil, New Zealand, Canada and the United States. In the United States, the commercial scale shale oil extraction began in 1857 at shale oil retorts retorting the Devonian oil shale along the Ohio River Valley. In Canada, the Craigleith Shale Oil Works started to retort oil shale of the Ordovician Whitby Formation near Collingwood, Ontario, on Lake Huron, in 1859. However, after crude oil discovery in Pennsylvania in 1859, the United States and Canadian oil shale industries found it difficult to compete and they were shut down by 1861.

In Australia, the first oil shale mine was commenced in 1865 at American Creek, Mount Kembla in New South Wales. At the same year, the first shale oil was produced by the Pioneer Kerosene Works at American Creek. A number of other mines were opened in the areas of the Great Dividing Range; however, by the turn of the 20th century they were closed because of the importation of cheaper crude oil. In Brazil, oil shale was first exploited in 1884 in Bahia.

In 1894, the Pumpherston retort (also known as the Bryson retort) was invented, which is considered as a separation of the oil shale industry from the coal industry. It stayed in use until 1938.

Operations during the 19th century focused on the production of kerosene, lamp oil, and paraffin; these products helped supply the growing demand for lighting that arose during the Industrial Revolution. Fuel oil, lubricating oil and grease, and ammonium sulfate were also produced.

### *First Half of the 20th Century*

The oil shale industry expanded immediately before World War I because of limited access to conventional petroleum resources and the mass production of automobiles and trucks, which accompanied an increase in gasoline consumption. The Office of Naval Petroleum and Oil Shale Reserves of the United States was established in 1912. The reserves were seen as a possible emergency source of fuel for the military, particularly the Navy. In 1915 an oil shale industry was established in Switzerland. The year 1916 is considered the beginning of the Estonian oil shale industry. In 1917, Russian paleontologist Mikhail Zalessky named kukersite oil shale after the Kukruse settlement. Continuous mining activities started shortly after. Initially, oil shale was used primarily in the cement industry, for firing in locomotive furnaces, and as a household fuel, followed by shale oil and power production. The first experimental oil shale processing retorts were built in 1921. In 1924, the Tallinn Power Plant was the first power plant in the world to employ oil shale as its primary fuel.

Between the World Wars oil shale projects were begun in Spain, China, Russia and South Africa; they restarted in Brazil and, for a short time, in Canada. In China, the extraction of oil shale began in 1926 under the Japanese rule. The commercial-scale production of shale oil began in 1930 in Fushun, Manchuria, with the construction of the "Refinery No. 1" operating Fushun-type retorts. In 1939–1945, a shale oil pilot plant operated in Morocco.

During World War II a modified *in situ* extraction process was used in Germany by digging tunnels into oil shale deposits and then collapsing their walls into the void volume to ignitate. This process had extremely low oil recovery and it was hard to control. In Sweden, the Swedish Shale Oil Company was formed in 1940. It exploited one of the earliest *in-situ* processes–underground gasification by electrical energy (Ljungström method)–between 1940 and 1966 at Kvarntorp. In 1940–1952, three NTU retorts were operated at Marangaroo, New South Wales, Australia.

### *1950s–1960s*

Although the Estonian, Russian and Chinese oil shale industries continued to grow after World War II, most other countries abandoned their projects due to high processing costs and the availability of cheaper petroleum. The shale oil extraction in Australia was discontinued in 1952 due to ceasing of government funding, in France in 1957, in Scotland and South Africa in 1962, and in Sweden and Spain in 1966. In Germany only Rohrbach Zement (now part of Holcim) in Dotternhausen continued using oil shale for cement, power and thermal energy production.

In 1946–1952, Dictyonema argillite was used for uranium production in Sillamäe. In Sweden the extraction of alum shale for uranium and vanadium continued until 1989. After World War II, Estonian-produced oil shale gas was used in Leningrad (Saint Petersburg) and in northern Estonia cities as a substitute for natural gas. In 1949, the Kohtla-Järve Power Plant – the first power plant in the world using pulverized oil shale at an industrial scale – was commissioned in Estonia. The world's two largest oil shale-fired power stations – Balti Power Plant and Eesti Power Plant (known as the Narva Power Plants) – were opened in 1965 and in 1973.

In China, the "Refinery No. 2" of Fushun began its production in 1954 and in 1959, the maximum annual shale oil production increased to 780,000 tonnes. The produced shale oil was used for producing light liquid fuels. In 1961, China was producing one third of its total oil production from oil shale.

The United States Bureau of Mines opened a demonstration mine at Anvils Point, west of Rifle, Colorado, which operated at a small scale. In the early 1960s TOSCO (The Oil Shale Corporation) opened an underground mine and built an experimental plant near Parachute, Colorado. It was closed in 1972 because the price of production exceeded the cost of imported crude oil. In 1951, the United States Department of Defence became interested in oil shale as an alternative resource for producing a jet fuel. In 1953, Sinclair Oil Corporation developed an *in situ* processing method using existing and induced fractures between vertical wells. In the 1960s, a proposal was suggested for a modified *in situ* process which involved creation of a rubble chimney (a zone in the rock formation created by breaking the rock into fragments) using a nuclear explosive.

### *1970s–1980s*

In 1972, the first modified *in situ* oil shale experiment in the United States was conducted by Occidental Petroleum at Logan Wash, Colorado. Due to the 1973 oil crisis, the oil shale industry restarted in several countries. The United States Navy and the Office of Naval Petroleum and Oil Shale Reserves started evaluations of oil shale's suitability for military fuels, such as jet fuels, marine fuels and a heavy fuel oil. Shale-oil based JP-4 jet fuel was produced until the early 1990s, when it was replaced with kerosene-based JP-8. In 1974 the United States Department of the Interior announced an oil shale leasing program in the oil shale regions of Colorado and Utah, and by the early 1980s almost all of the major oil companies had established oil shale pilot projects.

The United States oil shale industry collapsed when oil prices fell in the early 1980s. On 2 May 1982, known as "Black Sunday", Exxon canceled its US$5 billion Colony Shale Oil Project near Parachute,

Colorado because of low oil-prices and increased expenses, laying off more than 2,000 workers and leaving a trail of home-foreclosures and small-business bankruptcies. In 1986, President Ronald Reagan signed into law the Consolidated Omnibus Budget Reconciliation Act of 1985 which among other things abolished the United States' Synthetic Liquid Fuels Program. The last oil shale retort in the United States, operated by Unocal Corporation, closed in 1991.

Because of the success of oil shale-based power generation, Estonian oil shale production peaked in 1980 at 31.35 million tonnes. The largest oil shale mine in the world – the Estonia Mine – was opened in 1972. However, production decreased in Estonia during the 1990s, due to reduced demand from the power generation industry. It was mainly affected by construction of the nuclear power plants in the Soviet Union. Most of Russian oil shale mines were closed in the 1990s and production continued only on a small scale.

In Israel, a 0.1 MW pilot oil shale-fired power plant was tested in 1982–1986. A demonstration fluidised-bed plant became operational in 1989.

### Latest Developments

The global oil shale industry started to grow slightly in the mid-1990s. In 1992 commercial shale oil production using Petrosix technology resumed in Brazil. Estonian oil shale production has continuously increased since 1995. In 2005, China became the largest shale oil producer in the world. In Australia, a demonstration-scale processing plant at the Stuart Deposit near Gladstone, Queensland, produced between 2000 and 2004 over 1.5 million barrels ($240 \times 10^3$ $m^3$) of shale oil. In 2008–2009, the facility was dismantled. In the United States, an oil shale development program was initiated in 2003. The Energy Policy Act of 2005 introduced a commercial leasing program for oil shale and tar sands resources on public lands within the states of Colorado, Utah, and Wyoming.

In April 2010, the 4th Workshop on Regional Cooperation for Clean Utilization of Oil Shale was held in Egypt and later the same month an Oil Shale Cooperation Centre was established in Amman by Egypt, Jordan, Morocco, Syria and Turkey.

### Processing Principles

Shale oil extraction process decomposes oil shale and converts its kerogen into shale oil—a petroleum-like synthetic crude oil. The process is conducted by pyrolysis, hydrogenation, or thermal dissolution. The efficiencies of extraction processes are often evaluated by comparing their yields to the results of a Fischer Assay performed on a sample of the shale.

The oldest and the most common extraction method involves pyrolysis (also known as *retorting* or destructive distillation). In this process, oil shale is heated in the absence of oxygen until its kerogen decomposes into condensable shale oil vapours and non-condensable combustible oil shale gas. Oil vapours and oil shale gas are then collected and cooled, causing the shale oil to condense. In addition, oil shale processing produces spent oil shale, which is a solid residue. Spent shale consists of inorganic compounds (minerals) and char (some authors use the terms *coke residue* or *semi-coke* instead of *char*)—a carbonaceous residue formed from kerogen. Burning the char off the spent shale produces oil shale ash. Spent shale and shale ash can be used as ingredients in cement or brick manufacture. The composition of the oil shale may lend added value to the extraction process through the recovery of by-products, including ammonia, sulfur, aromatic compounds, pitch, asphalt, and waxes.

Heating the oil shale to pyrolysis temperature and completing the endothermic kerogen decomposition reactions require a source of energy. Some technologies burn other fossil fuels such as natural gas, oil, or coal to generate this heat and experimental methods have used electricity, radio waves, microwaves, or reactive fluids for this purpose. Two strategies are used to reduce, and even eliminate, external heat energy requirements: the oil shale gas and char by-products generated by pyrolysis may be burned as a source of energy, and the heat contained in hot spent oil shale and oil shale ash may be used to pre-heat the raw oil shale.

For *ex situ* processing, oil shale is crushed into smaller pieces, increasing surface area for better extraction. The temperature at which decomposition of oil shale occurs depends on the time-scale of the process. In *ex situ* retorting processes, it begins at 300 °C (570 °F) and proceeds more rapidly and completely at higher temperatures. The amount of oil produced is the highest when the temperature ranges between 480 and 520 °C (900 and 970 °F). The ratio of oil shale gas to shale oil generally increases along with retorting temperatures. For a modern *in situ* process, which might take several months of heating, decomposition may be conducted at temperatures as low as 250 °C (480 °F).

Temperatures below 600 °C (1,110 °F) are preferable, as this prevents the decomposition of lime stone and dolomite in the rock and thereby limits carbon dioxide emissions and energy consumption.

Hydrogenation and thermal dissolution (reactive fluid processes) extract the oil using hydrogen donors, solvents, or a combination of these. Thermal dissolution involves the application of solvents at elevated temperatures and pressures, increasing oil output by cracking the dissolved organic matter. Different methods produce shale oil with different properties.

### Classification of Extraction Technologies

Industry analysts have created several classifications of the technologies used to extract shale oil from oil shale.

***By process principles*:** Based on the treatment of raw oil shale by heat and solvents the methods are classified as pyrolysis, hydrogenation, or thermal dissolution.

***By location*:** A frequently used distinction considers whether processing is done above or below ground, and classifies the technologies broadly as *ex situ* (displaced) or *in situ* (in place). In *ex situ* processing, also known as above-ground retorting, the oil shale is mined either underground or at the surface and then transported to a processing facility. In contrast, *in situ* processing converts the kerogen while it is still in the form of an oil shale deposit, following which it is then extracted via oil wells, where it rises in the same way as conventional crude oil. Unlike *ex situ* processing, it does not involve mining or spent oil shale disposal aboveground as spent oil shale stays underground.

***By heating method*:** The method of transfering heat from combustion products to the oil shale may be classified as direct or indirect. While methods that allow combustion products to contact the oil shale within the retort are classified as *direct*, methods that burn materials external to the retort to heat another material that contacts the oil shale are described as *indirect*

***By heat carrier*:** Based on the material used to deliver heat energy to the oil shale, processing technologies have been classified into gas heat carrier, solid heat carrier, wall conduction, reactive fluid, and volumetric heating methods. Heat carrier methods can be sub-classified as direct or indirect.

***By raw oil shale particle size*:** The various *ex situ* processing technologies may be differentiated by the size of the oil shale particles that are fed into the retorts. As a rule, gas heat carrier technologies process oil shale lumps varying in diameter from 10 to 100 millimetres (0.4 to 3.9 in), while solid heat carrier and wall conduction technologies process fines which are particles less than 10 millimetres (0.4 in) in diameter.

***By retort orientation*:** "Ex-situ" technologies are sometimes classified as vertical or horizontal. Vertical retorts are usually shaft kilns where a bed of shale moves from top to bottom by gravity. Horizontal retorts are usually horizontal rotating drums or screws where shale moves from one end to the other. As a general rule, vertical retorts process lumps using a gas heat carrier, while horizontal retorts process fines using solid heat carrier.

***By complexity of technology*:** *In situ* technologies are usually classified either as *true in situ* processes or *modified in situ* processes. *True in situ* processes do not involve mining or crushing the oil shale. *Modified in situ* processes involve drilling and fracturing the target oil shale deposit to create voids in the deposit. The voids enable a better flow of gases and fluids through the deposit, thereby increasing the volume and quality of the shale oil produced.

# Bibliography


Anderson, Irvine H.: *Aramco, the United States, and Saudi Arabia: A Study of the Dynamics of Foreign Oil Policy*, 1933-1950. Princeton: Princeton University Press, 1981.

Armistead, George: *Safety in Petroleum Refining and Related Industries*, New York: John G. Simmons & Co., 1950.

Balakrishan, A. : *Biofuel Crops and Renewable Energy*, Sublime Publications, Delhi, 2011.

Bhattacharya, K. G. and A. K. Talukdar: *Catalysis in Petroleum and Petrochemical Industries*, Alpha Science International, Delhi, 2005.

Black, Brian: *Petrolia: The Landscape of America's First Oil Boom*. Baltimore: Johns Hopkins University Press, 2000.

Blomen, Leo, and Michael Mugerwa: *Fuel Cell Systems*. New York: Plenum Press, 1993.

Bromley, Simon: *American Hegemony and World Oil: The Industry, the State System, and the World Economy*. University Park, Pa., 1991.

Burdick, Donald L., and William L. Leffler: *Petrochemicals in Nontechnical Language*. Tulsa, Okla.: PennWell, 2001.

Chester, Edward W.: *United States Oil Policy and Diplomacy: A Twentieth-Century Overview*. West-port, Conn.: Greenwood Press, 1983.

Chisholm, Archibald H. T.: *The First Kuwait Oil Concession Agreement: A Record of the Negotiations*, 1911-1934. London: Frank Cass, 1975.

Chowdhary, L.R. : *Petroleum Geology of the Cambay Basin, Gujarat, India*, Indian Petroleum, Delhi, 2004.

Clark, John G.: *The Political Economy of World Energy: A Twentieth-Century Perspective*. Chapel Hill, N.C., 1990.

Connelly, William L.: *The Oil Business As I Saw It: Half a Century with Sinclair*. Norman: University of Oklahoma Press, 1954.

Cordesman, Anthony H.: *The Gulf and the West: Strategic Relations and Military Realities*. Boulder, Colo.: Westview Press, 1988.

Cowhey, Peter F.: *The Problems of Plenty: Energy Policy and International Politics*. Berkeley: Uni-versity of California Press, 1985.

Cutler, Tom: *The Military Demand for Oil*. Petroleum Economist, London, 1989.

Davenport, E. H. and Sidney Russell Cooke: *The Oil Trusts and Anglo-American Relations.* Macmillan, 1924

DeLaisi, Francis: *Oil, Its Influence on Politics.* Labor Publishing,1923.

Diwan, Parag and Ashutosh Karnatak: *Handbook of Natural Gas Technology and Business*, Pentagon Energy Press, Delhi, 2009.

Eaton, Rev. S.J.M.: *Petroleum, a History of the Oil Region*, J.P. Skelly & Co., Philadelphia, 1866.

Engler, Robert: *The Politics of Oil: A Study of Private Power and Democratic Directions.* New York, Macmillan, 1961.

Feis, Herbert: *Petroleum and American Foreign Policy.* Stanford: Food Research Institute, 1944.

Fischer, Louis: *Oil Imperialism: The International Struggle for Petroleum.* New York: International Publishers, 1926.

Frankel, Paul: *Mattei: Oil and Power Politics.* New York and Washington: Praeger, 1966.

Garverick, Linda: *Corrosion in the Petrochemical Industry.* Materials Park, Ohio: ASM International, 2000.

Giddens, Paul H.: *The Birth of the Oil Industry*, The MacMillian Company, New York, 1938.

Gramling, Robert: *Oil on the Edge: Offshore Development, Conflict, Gridlock.* Albany: State University of New York Press, 1996.

Hamilton, Adrian: *Oil: The Price of Power.* London: Michael Joseph/Rainbird, 1986.

Hartshorn, J. E.: *Oil Companies and Governments: An Account of the International Oil Industry in Its Political Environment.* London: Faber and Faber, 1962.

Hyne, Norman J.: *Nontechnical Guide to Petroleum Geology, Exploration, Drilling, and Production.* PennWell Corporation, Delhi, 2001.

Jentleson, Bruce: *Pipeline Politics: The Complex Political Economy of East-West Energy Trade.* lthaca: Cornell University Press, 1986.

Kazemzadeh, Firuz: *The Struggle for Transcaucasia, 1917-1921.* Philosophical Library, 1951.

Kent, Marian: *Oil and Empire: British Policy and Mesopotamian Oil, 1900-1920.* London: Macmillan, 1976.

Klare, Michael: *Blood and Oil: The Dangers and Consequences of America's Growing Petroleum Diplomacy.* Metropolitan Books/ Henry Holt, 2004.

Klare, Michael: *Resource Wars: The New Landscape of Global Conflict.* Metropolitan Books / Henry Holt, 2001.

Kleveman, Lutz: *The New Great Game: Blood and Oil in Central Asia.* Atlantic Books, 2003.

Kordesch, Karl, and Günter Simader: *Fuel Cells and Their Applications.* New York: VCH, 1996.

Lieber, Robert J.: *Oil and the Middle East War: Europe in the Energy Crisis.* Cambridge: Center for International Affairs, 1976.

Lubell, Harold: *Middle East Oil Crises and Western Europe's Energy Supplies.* Baltimore: Johns Hopkins University Press, 1963.

Marsh, Steve: *Anglo-American Relations and Cold War Oil: Crisis in Iran.* New York: Palgrave MacMillan, 2003.

McLaurin, John J.: *Sketches in Crude-oil*, J. Horace McFarland Company, Harrisburg, 1896.

McNaugher, Thomas: *Arms and Oil: US Military Strategy and the Persian Gulf.* Brookings Institution, Londong, 1985.

McQuaig, Linda. *It's The Crude, Dude: War Big Oil, And The Fight For The Planet.* Doubleday Canada, 2004.

Mejcher, Helmut. *Imperial Quest for Oil: Iraq, 1910-1928.* Ithaca Press, NY, 1976.

Meyer, Lorenzo: *Mexico and the United States in the Oil Controversy: 1917-1942.* University of Texas Press, New York. 1977.

Miller, Aaron David: *Search for Security: Saudi Arabian Oil and American Foreign Policy, 1939-1949.* Chapel Hill: University of North Carolina Press, 1080.

Mohr, Anton. *The Oil War.* Harcourt, Brace, 1926. Hyperion Press, 1976.

Mosley, Leonard: *Power Play: Oil in the Middle East.* New York: Random House, 1973.

Naderpour, N and F Hematti Kordmahaleh: *Petroleum Geosciences*, SBS Pub, Delhi, 2009.

O'Connor, Harvey *The Empire of Oil.* Monthly Review Press, 1955

Odell, Peter R.: *Oil and World Power: Background of the Oil Crisis,* New York: Viking Penguin, 1986.

Olien, Diana Davids: *Oil in Texas: The Gusher Age, 1895–1945.* Austin: University of Texas Press, 2002.

Olien, Roger: *Oil and Ideology: The Cultural Creation of the American Petroleum Industry.* Chapel Hill: University of North Carolina Press, 2000.

Penrose, Edith, and E. F. Penrose: *Iraq: International Relations and National Development.* London: Ernest Bern, 1978.

Philip, George: *Oil and Politics in Latin America: Nationalist Movements and State Companies.* Cambridge University Press, Cambridge, 1982.

Randall, Stephen J. *United States Foreign Oil Policy, 1919–1948: For Profits and Security.* Kingston, Ontario, 1985.

Rathi, Rakesh : *Petroleum Refining Processes*, SBS Pub, Delhi, 2007.

Redlinger, Robert Y.: *Wind Energy in the 21st Century: Economics, Policy, Technology, and the Changing Electricity Industry.* New York: Macmillan, 2001.

Rister, Carl Coke: *Oil! Titan of the Southwest.* Norman: University of Oklahoma Press, 1949.

Sah, S.L. : *Encyclopaedia of Petroleum Science and Engineering*, Kalpaz, Delhi, 2003.

Sampson, Anthony: *The Seven Sisters: The Great Oil Companies and the World They Shaped.* New York, 1975.

Satyanarayana, D. : *Petroleum Geochemistry*, Daya Publishing, Delhi, 2011.

Schneider, Steven A.: *The Oil Price Revolution.* Baltimore, 1983.

Simmons, Matthew R.: *Twlight in the Desert: The Coming Saudi Oil Shock and the World Economy.* 2005

Singh, Lakshman and Anoop Tripathi: *Indian Petroleum Directory : With an Introductory Chapter on Petroleum Industry in India*, Indian Petroleum Pub, Delhi, 2007.

Speight, James G.: *The Chemistry and Terminology of Petroleum*. New York: Marcel Dekker, 1980.

Srivastava, Rajesh K. , Ch. Sivaji and N.V. Chalapathi Rao: *Indian Dykes: Geochemistry Geophysics and Geochronology*, Narosa, Delhi, 2008.

Stoff, Michael B. *Oil, War, and American Security*. Yale University Press, 1980.

Subramaniam, A P and S Balakrishna: *Advancing Frontiers in Geology and Geophysics,* Today and Tomorrow's, Delhi, 2007.

Tiwari, S.K. : *Fuel Minerals and Other Energy Resources*, Atlantic Pub, Delhi, 2010.

Tribedi, S. : *Proceedings of National Convention of Environment in India, Challenge for the 21st Century*, India, 1994.

Venn, Fiona: *Oil Diplomacy in the Twentieth Century.* New York, 1986. Pioneering study stressing oil's strategic importance.

Vervalin, Charles H.: *Fire Protection Manual for Hydrocarbon Processing Plants*. Houston: Gulf Publishing Co, 1985.

Vijayalakshmi, G S : *Fuels and Biofuels*, Agrobios, Delhi, 2007.

Watkins, Eric: *The Middle Eastern Environment*, London, The British Society for Middle Eastern Studie, 1995.

Wendel, Charles H.: *The Allis-Chalmers Story*. Sarasota, Florida: Crestline Publishing Co., 1988.

Yergin, Daniel: *The Prize: The Epic Quest for Oil, Money, and Power.* New York: Simon and Schuster, 1991.

Young, George J.: *Fuel Cells,* New York: Reinhold Publishing Corp., 1961.

# Index

□□□